W9-ARM-844

STEPHEN BIESTY'S INCREDIBLE EXPLOSIONS

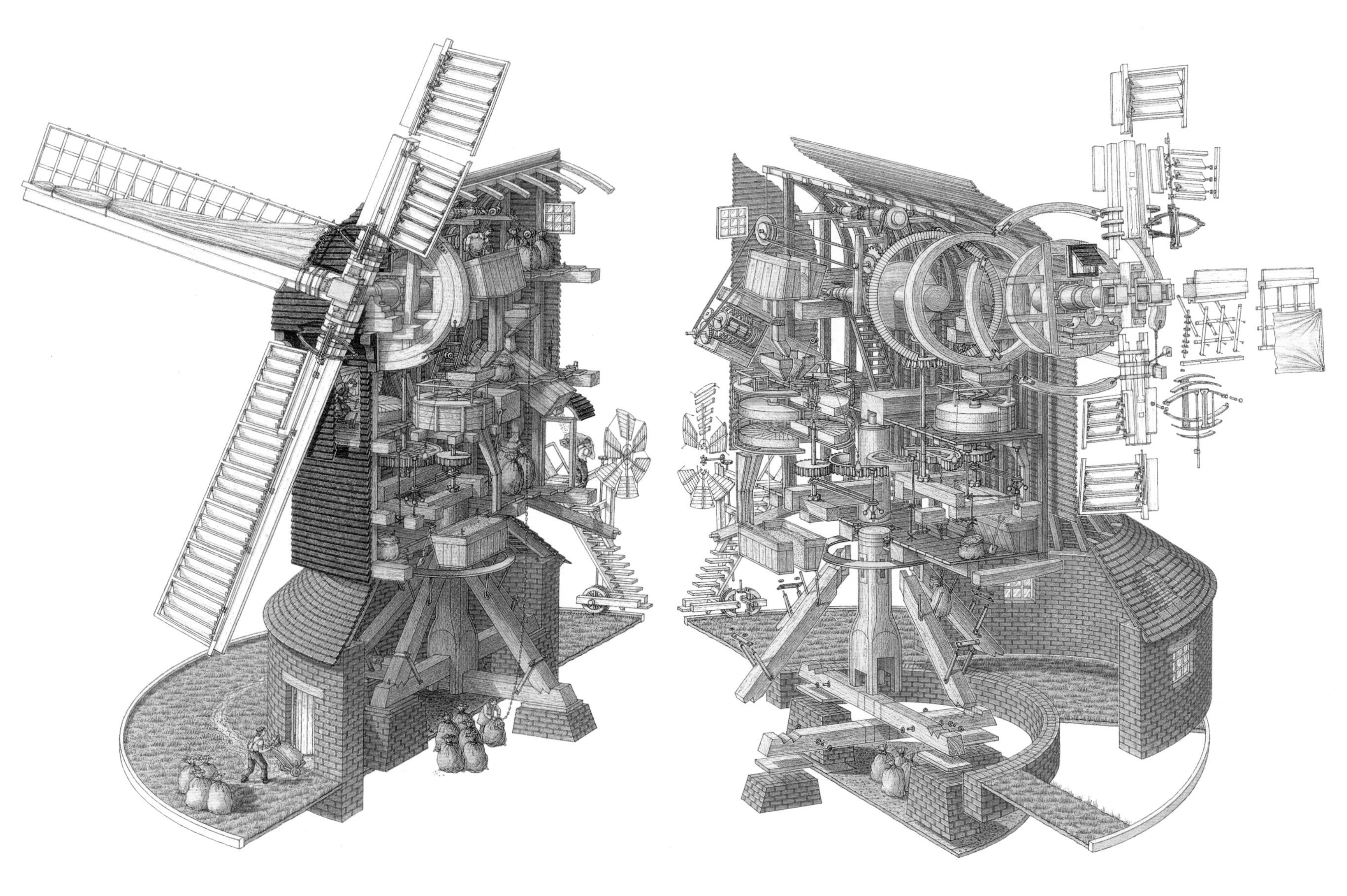

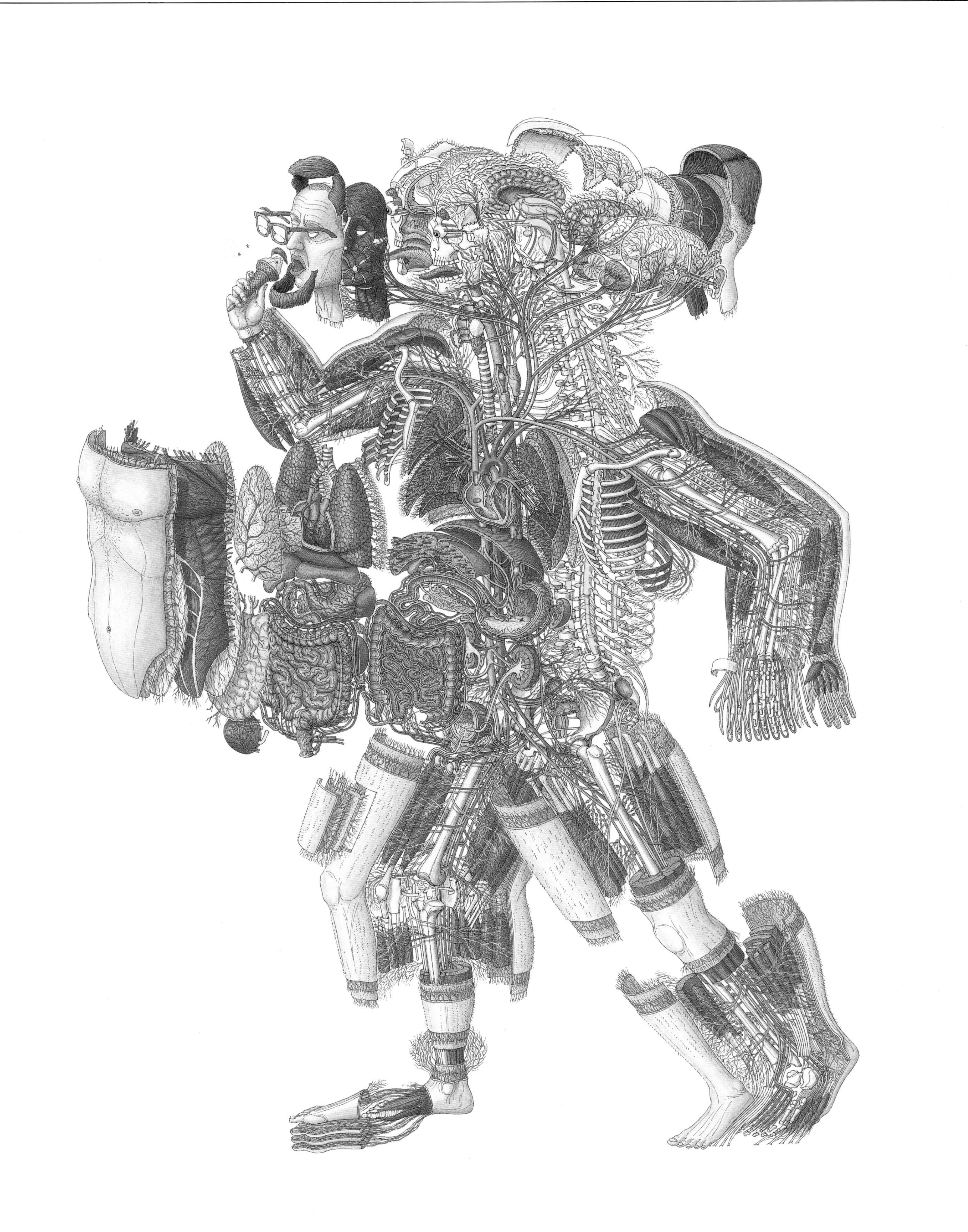

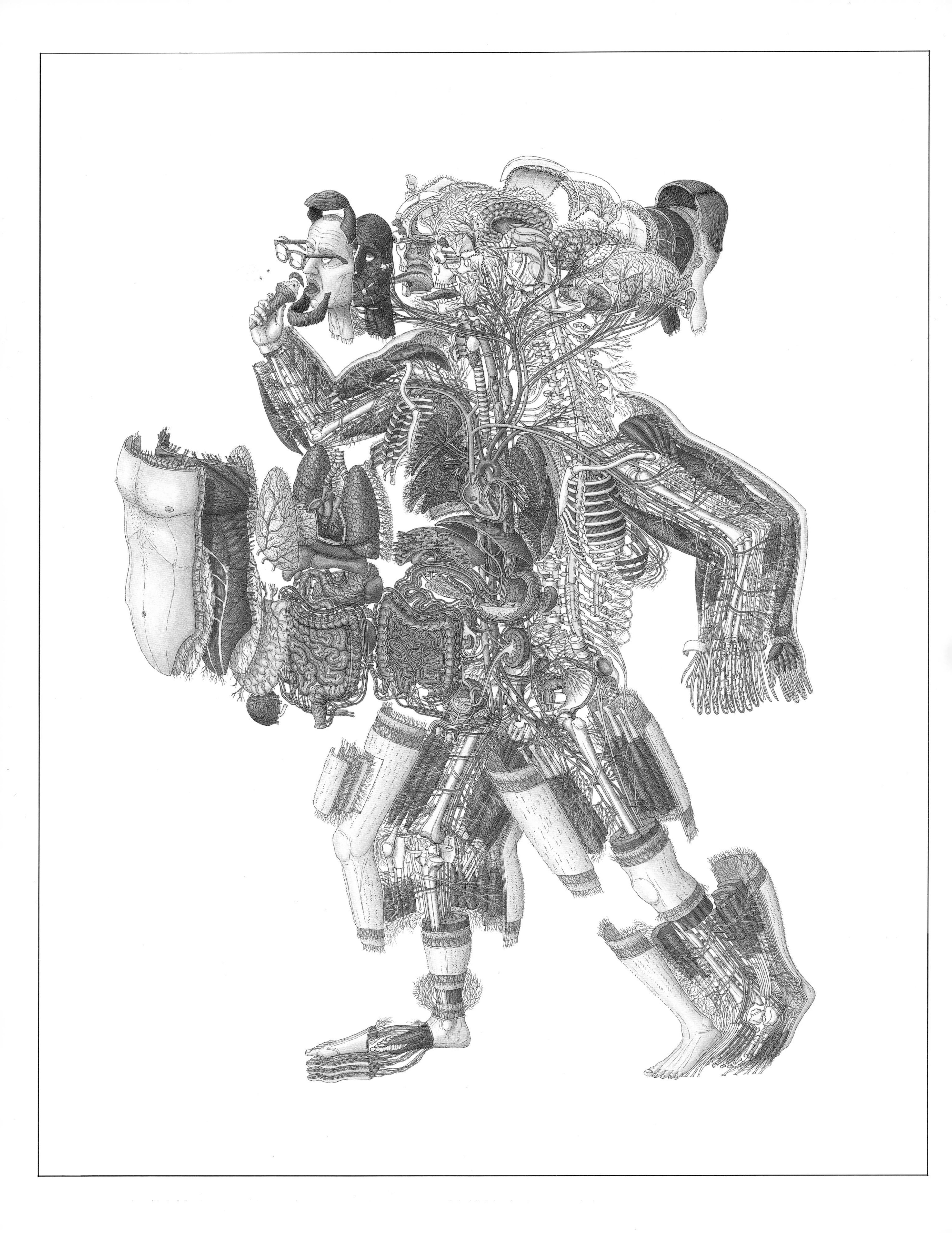

STEPHEN BIESTY'S INCREDIBLE EXPLOSIONS

ILLUSTRATED BY
STEPHEN BIESTY

WRITTEN BY
RICHARD PLATT

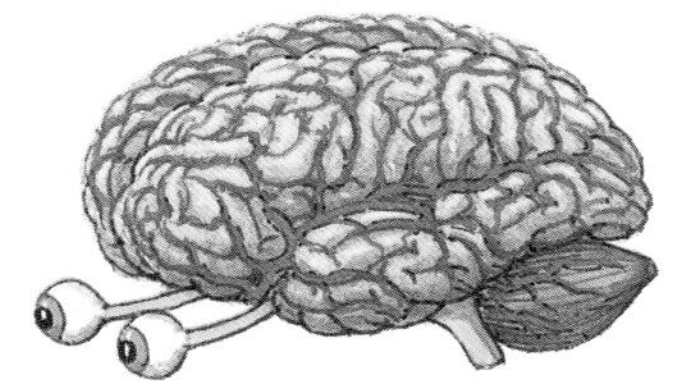

A DK PUBLISHING BOOK

Art Editors Dorian Spencer Davies, Sharon Grant
Senior Art Editor C. David Gillingwater
U.S. Editor Camela Decaire
Senior Editor John C. Miles
Production Louise Barratt

First American edition, 1996
2 4 6 8 10 9 7 5 3 1
Published in the United States
by DK Publishing, Inc.,
95 Madison Avenue, New York, New York 10016

A CIP catalog record is available
from the Library of Congress

ISBN: 0-7894-1024-9

Greetings, Earthlings!

The atomic drive and date controller on my space cruiser have broken down. Now, when I try to land on your planet, I keep crashing at the wrong place and time. It's annoying for me, but it is a great way to meet Earth people. You have all made me very welcome: artist Stephen Biesty helped me repair my craft when I crashed near his home. I told him of all the other places I had seen, and he drew them for me as a souvenir of my trip to Earth so far. Perhaps you can spot my crash landing sites – there is one in each picture in this book.

Reproduced by Dot Gradations, Essex
Printed in Italy by L.E.G.O.

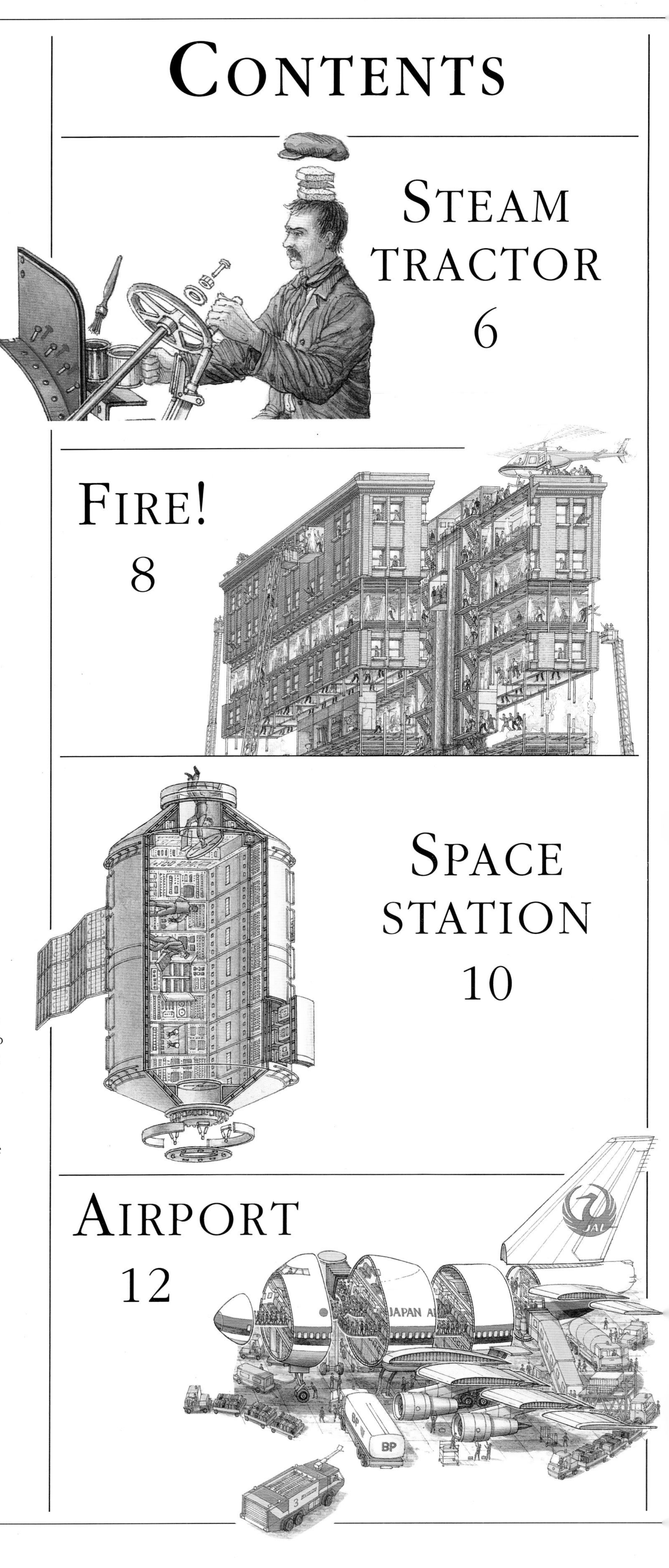

Contents

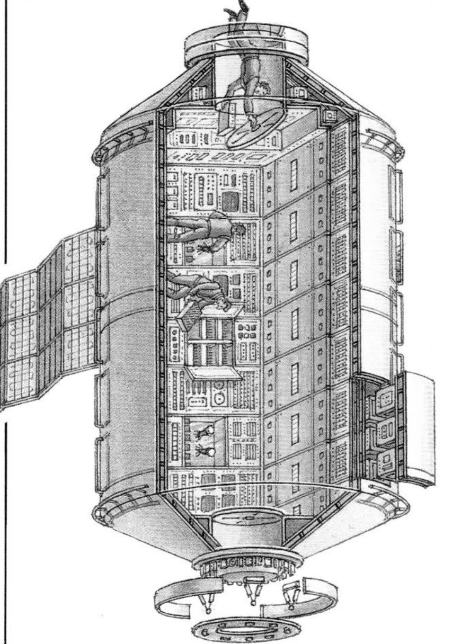

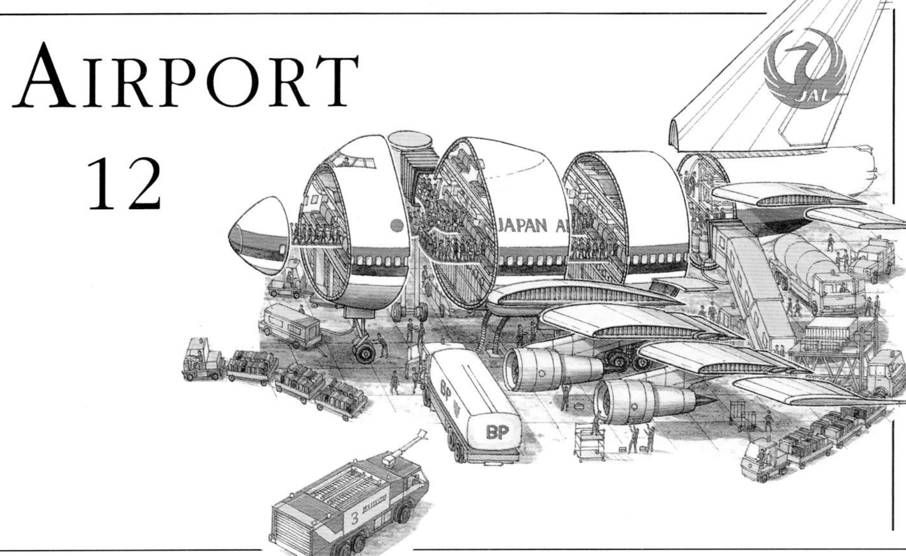

Steam tractor

HISSING, STEAMING, CLANKING, AND SHAKING THE ground, steam tractors are like mechanical dinosaurs. Now almost extinct, they once embodied the most modern form of power, the steam engine. Steam tractors first appeared around 1860. At first they powered stationary farm machinery, such as threshers, which separate kernels of grain from their husks. Steam tractors then pulled plows or heavy road wagons, and turned fairground rides. For a while, steam represented a bright, labor-saving future. But the reign of the steam engine was a short one. By the beginning of this century, electric motors and internal combustion engines had begun to take over many of its tasks. Before long, even the puffing steam tractors were making their last journeys – to the scrap yard or museum.

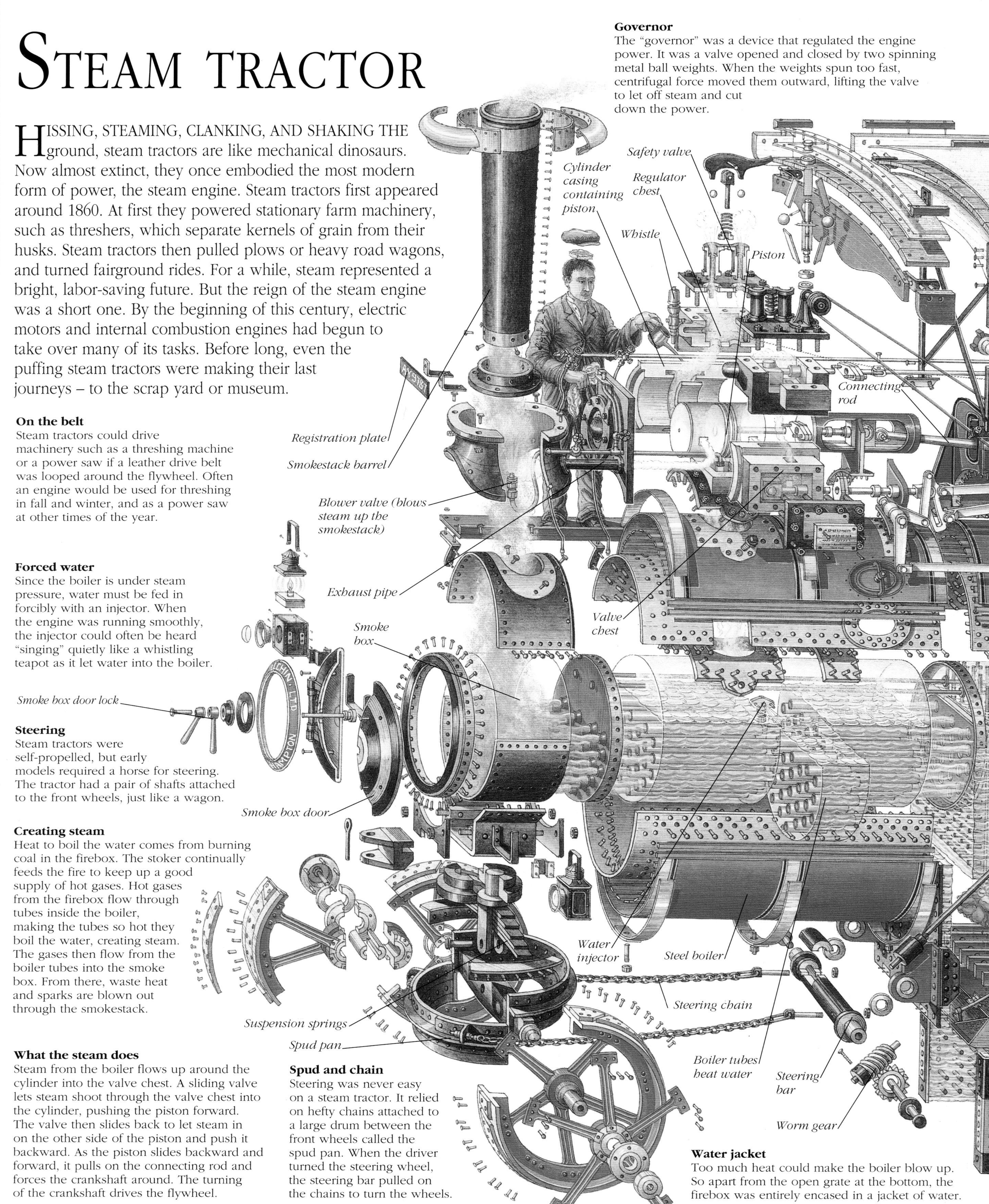

Governor
The "governor" was a device that regulated the engine power. It was a valve opened and closed by two spinning metal ball weights. When the weights spun too fast, centrifugal force moved them outward, lifting the valve to let off steam and cut down the power.

On the belt
Steam tractors could drive machinery such as a threshing machine or a power saw if a leather drive belt was looped around the flywheel. Often an engine would be used for threshing in fall and winter, and as a power saw at other times of the year.

Forced water
Since the boiler is under steam pressure, water must be fed in forcibly with an injector. When the engine was running smoothly, the injector could often be heard "singing" quietly like a whistling teapot as it let water into the boiler.

Steering
Steam tractors were self-propelled, but early models required a horse for steering. The tractor had a pair of shafts attached to the front wheels, just like a wagon.

Creating steam
Heat to boil the water comes from burning coal in the firebox. The stoker continually feeds the fire to keep up a good supply of hot gases. Hot gases from the firebox flow through tubes inside the boiler, making the tubes so hot they boil the water, creating steam. The gases then flow from the boiler tubes into the smoke box. From there, waste heat and sparks are blown out through the smokestack.

What the steam does
Steam from the boiler flows up around the cylinder into the valve chest. A sliding valve lets steam shoot through the valve chest into the cylinder, pushing the piston forward. The valve then slides back to let steam in on the other side of the piston and push it backward. As the piston slides backward and forward, it pulls on the connecting rod and forces the crankshaft around. The turning of the crankshaft drives the flywheel.

Spud and chain
Steering was never easy on a steam tractor. It relied on hefty chains attached to a large drum between the front wheels called the spud pan. When the driver turned the steering wheel, the steering bar pulled on the chains to turn the wheels.

Water jacket
Too much heat could make the boiler blow up. So apart from the open grate at the bottom, the firebox was entirely encased in a jacket of water.

Blasted boiler
Boiler explosions were a frequent cause of serious accidents. They usually happened when a tractor driver interfered with the safety valve to increase speed. Among the best known was an 1880 explosion in the English town of Maidstone. The explosion killed the driver's assistant and blasted the whole neighborhood with bits of the trailer and its load – sewage.

Engine driver
Steam tractor drivers moved around from village to village, following the work. They had the reputation of being dishonest, rowdy drunks. They earned 10% more than the best-paid farm worker, so they could afford to drink heavily.

The crew
Typically, a steam tractor had a crew of three – steersman or driver, stoker, and oil boy. But sometimes extra men were needed to run special machinery or tend a trailer.

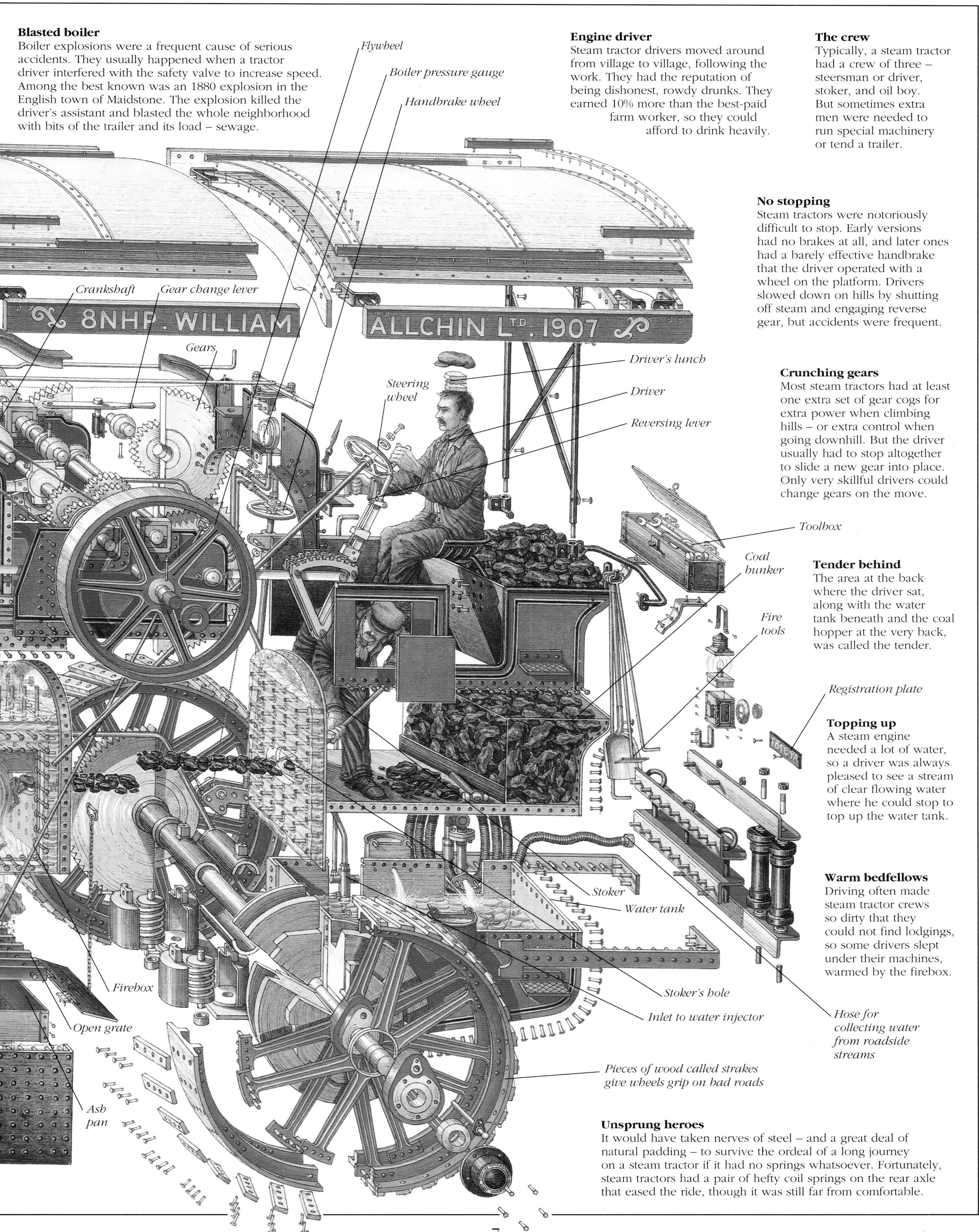

No stopping
Steam tractors were notoriously difficult to stop. Early versions had no brakes at all, and later ones had a barely effective handbrake that the driver operated with a wheel on the platform. Drivers slowed down on hills by shutting off steam and engaging reverse gear, but accidents were frequent.

Crunching gears
Most steam tractors had at least one extra set of gear cogs for extra power when climbing hills – or extra control when going downhill. But the driver usually had to stop altogether to slide a new gear into place. Only very skillful drivers could change gears on the move.

Tender behind
The area at the back where the driver sat, along with the water tank beneath and the coal hopper at the very back, was called the tender.

Topping up
A steam engine needed a lot of water, so a driver was always pleased to see a stream of clear flowing water where he could stop to top up the water tank.

Warm bedfellows
Driving often made steam tractor crews so dirty that they could not find lodgings, so some drivers slept under their machines, warmed by the firebox.

Unsprung heroes
It would have taken nerves of steel – and a great deal of natural padding – to survive the ordeal of a long journey on a steam tractor if it had no springs whatsoever. Fortunately, steam tractors had a pair of hefty coil springs on the rear axle that eased the ride, though it was still far from comfortable.

Fire!

Burning in a city high-rise, fire is like a dangerous monster. It attacks your senses one by one. You can often smell the hot-tempered beast before you see it. As a fire grows by consuming everything in its path, you hear its voice, first crackling, then roaring, in your ears. For city dwellers, the fire monster is an old enemy, and one that has never been far away. Today we're learning how to control the monster and curb its appetite. At home, simple safety precautions such as inexpensive smoke detectors warn of danger. If flames take hold, a 911 call brings brave firefighters rushing to the blaze. Water from their hoses quickly cools the flames and creates steam that starves the fire of the air it needs to burn.

Spreading smoke
An elevator shaft can channel choking smoke everywhere. Newer buildings now incorporate pressurizing fans that force air into the shaft, keeping it free of smoke.

Helicopter rescues victims

Sprinkler

Chopper rescue
When the tallest buildings catch fire, a lift from a helicopter may be the only escape for those trapped on the upper floors.

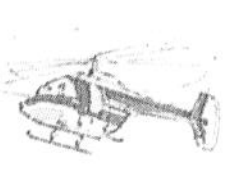

A quiet read
Fire alarms should alert everyone in the building to the danger. Nevertheless, experience shows that when there are no other signs of fire, some people don't recognize the alarm signal, or they just prefer to ignore it.

In the dark
In a smoke-filled room, visibility is zero, and firefighters have to navigate by touch alone. As they search, they lay down a cord so that they can retrace their steps to safety.

Aerial platform rescues people

Aerial ladder

Don't panic
Recent research suggests that fire rarely causes panic. Though aware of the danger, people usually help each other escape. A stampede to safety happens only when victims believe the fire is about to cut off their escape route.

Sprinklers at work
By spraying a fire with water, automatic sprinklers can put out flames before they take hold. Room temperatures much higher than 140°F (60°C) fracture a soft metal strip or a tiny glass bulb on the sprinkler nozzle; water then sprays out, covering about 130 sq ft (12 sq m).

Smoke rises through ventilating shafts

Anybody in?
Firefighters must search every room to check that there's nobody unconscious, asleep, or perhaps too old or ill to move.

Save my pianos!
Automatic sprinklers are not new. Henry S. Parmelee of Connecticut invented them in 1875 to protect his piano factory from fire. New England cotton mills made them a success – the fibers they processed made the mills a big fire risk, and without a sprinkler system, mill owners couldn't get fire insurance.

Fetching hoses
Elevator shafts in new buildings are protected against fire, so firefighters use them to rush hoses to floors where they are needed.

Hydraulic platform

Bad buttons
Some elevator buttons use heat sensors to summon the elevator. In a fire, they're fatal – the elevator goes to the burning floor.

A long way down
The tallest turntable ladders reach floors up to 165 ft (50 m) above the ground.

Hydraulic snorkel

Wet risers
In most nations, tall buildings must have wet risers – pipes that channel water to hydrants on each floor. Firefighters connect hoses directly to the wet risers.

Fireboat
When a dockside building bursts into flames, fire authorities can call on the city fireboat to help. It more often sprays water or foam on burning ships.

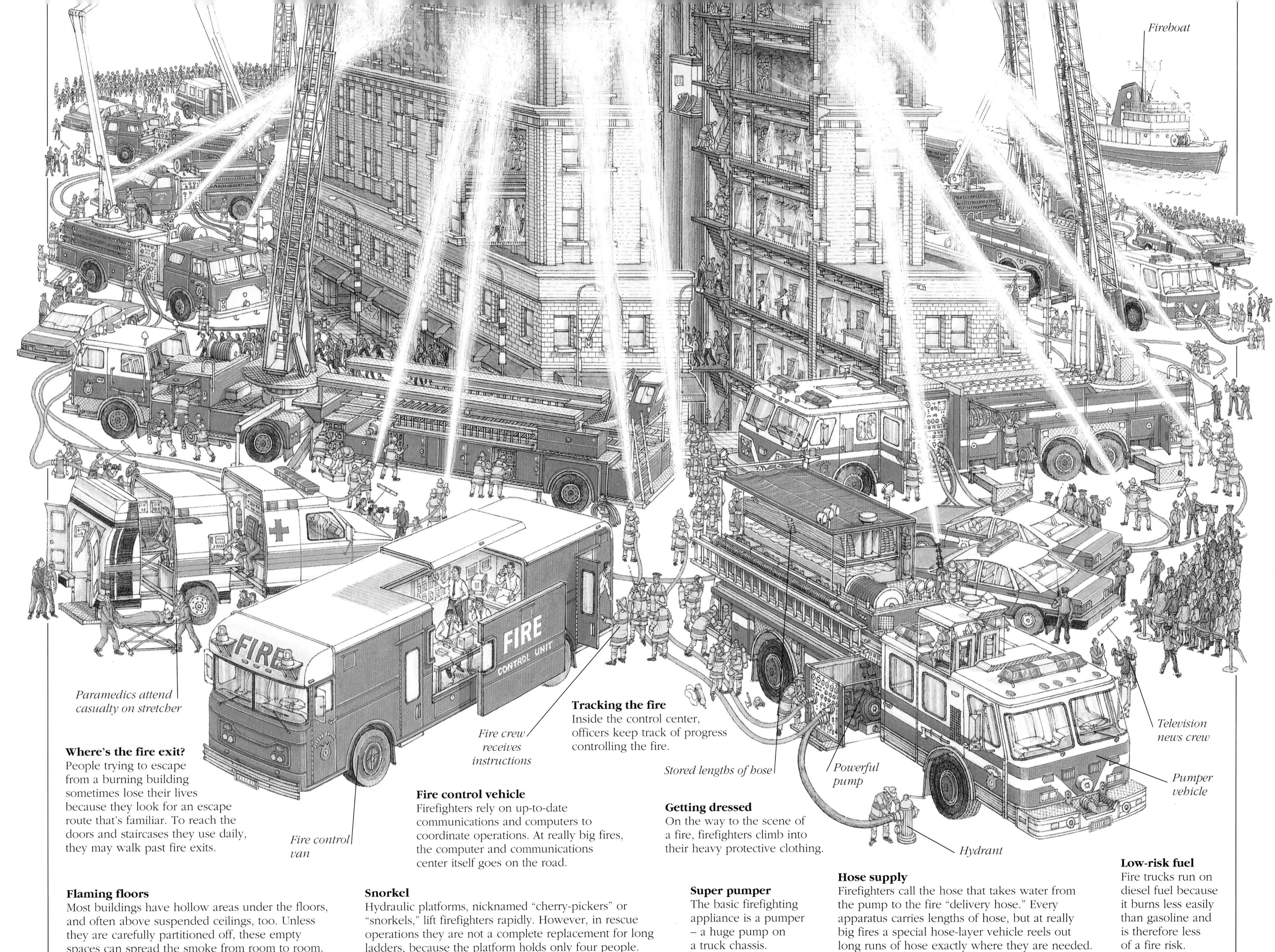

Paramedics attend casualty on stretcher

Fire control van

Fire crew receives instructions

Stored lengths of hose

Powerful pump

Hydrant

Pumper vehicle

Television news crew

Where's the fire exit?
People trying to escape from a burning building sometimes lose their lives because they look for an escape route that's familiar. To reach the doors and staircases they use daily, they may walk past fire exits.

Fire control vehicle
Firefighters rely on up-to-date communications and computers to coordinate operations. At really big fires, the computer and communications center itself goes on the road.

Tracking the fire
Inside the control center, officers keep track of progress controlling the fire.

Getting dressed
On the way to the scene of a fire, firefighters climb into their heavy protective clothing.

Flaming floors
Most buildings have hollow areas under the floors, and often above suspended ceilings, too. Unless they are carefully partitioned off, these empty spaces can spread the smoke from room to room.

Snorkel
Hydraulic platforms, nicknamed "cherry-pickers" or "snorkels," lift firefighters rapidly. However, in rescue operations they are not a complete replacement for long ladders, because the platform holds only four people.

Super pumper
The basic firefighting appliance is a pumper – a huge pump on a truck chassis.

Hose supply
Firefighters call the hose that takes water from the pump to the fire "delivery hose." Every apparatus carries lengths of hose, but at really big fires a special hose-layer vehicle reels out long runs of hose exactly where they are needed.

Low-risk fuel
Fire trucks run on diesel fuel because it burns less easily than gasoline and is therefore less of a fire risk.

Space station

Satellite capture

One of the main uses of the space station is the repair of damaged satellites. The crew uses the remote manipulator arm to capture the satellite, or to transfer it from the *Teleoperator* robot, collecting satellites from distant orbits.

WHICH WAY IS UP? HERE ON EARTH, THE ANSWER IS EASY. BUT TO AN ASTRONAUT on board a space station, it's a silly question. "Standing up" or "putting down a book" doesn't mean very much when everything is weightless. We get a sense of up and down from gravity. Scientists call the weightlessness experienced in space "microgravity." They plan to use it to make ultrapure crystals or exotic alloys (mixtures) of metals. Microgravity can be a problem, though. Walking is impossible when there's no gravity to press your feet firmly against the "floor." Drinking is tough when the coffee floats around. American astronauts and Russian cosmonauts have found ways around these difficulties and the two nations hope to launch a joint space station soon. Plans have changed many times, but the joint craft may look roughly like this.

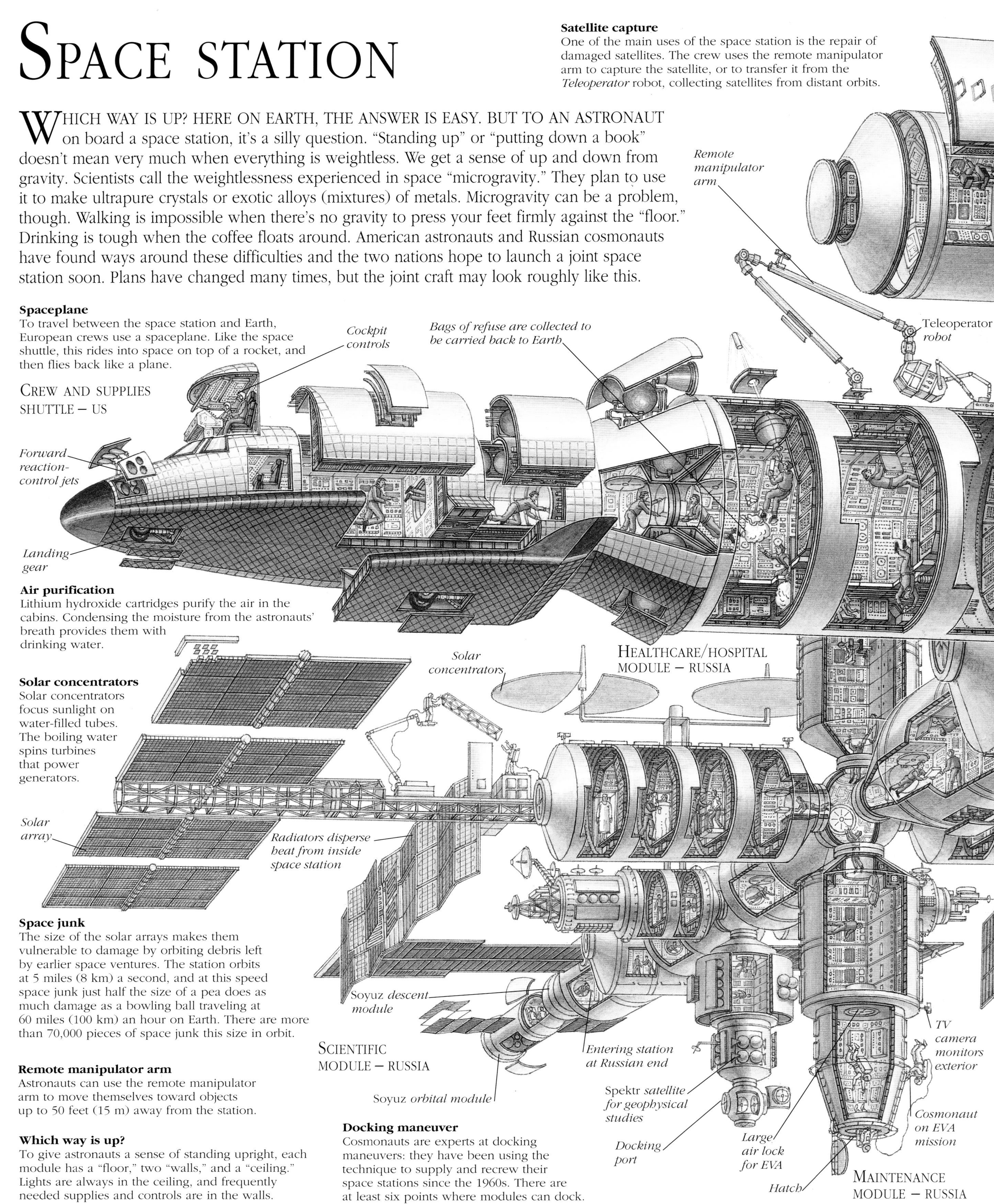

Spaceplane

To travel between the space station and Earth, European crews use a spaceplane. Like the space shuttle, this rides into space on top of a rocket, and then flies back like a plane.

Air purification

Lithium hydroxide cartridges purify the air in the cabins. Condensing the moisture from the astronauts' breath provides them with drinking water.

Solar concentrators

Solar concentrators focus sunlight on water-filled tubes. The boiling water spins turbines that power generators.

Space junk

The size of the solar arrays makes them vulnerable to damage by orbiting debris left by earlier space ventures. The station orbits at 5 miles (8 km) a second, and at this speed space junk just half the size of a pea does as much damage as a bowling ball traveling at 60 miles (100 km) an hour on Earth. There are more than 70,000 pieces of space junk this size in orbit.

Remote manipulator arm

Astronauts can use the remote manipulator arm to move themselves toward objects up to 50 feet (15 m) away from the station.

Which way is up?

To give astronauts a sense of standing upright, each module has a "floor," two "walls," and a "ceiling." Lights are always in the ceiling, and frequently needed supplies and controls are in the walls.

Docking maneuver

Cosmonauts are experts at docking maneuvers: they have been using the technique to supply and recrew their space stations since the 1960s. There are at least six points where modules can dock.

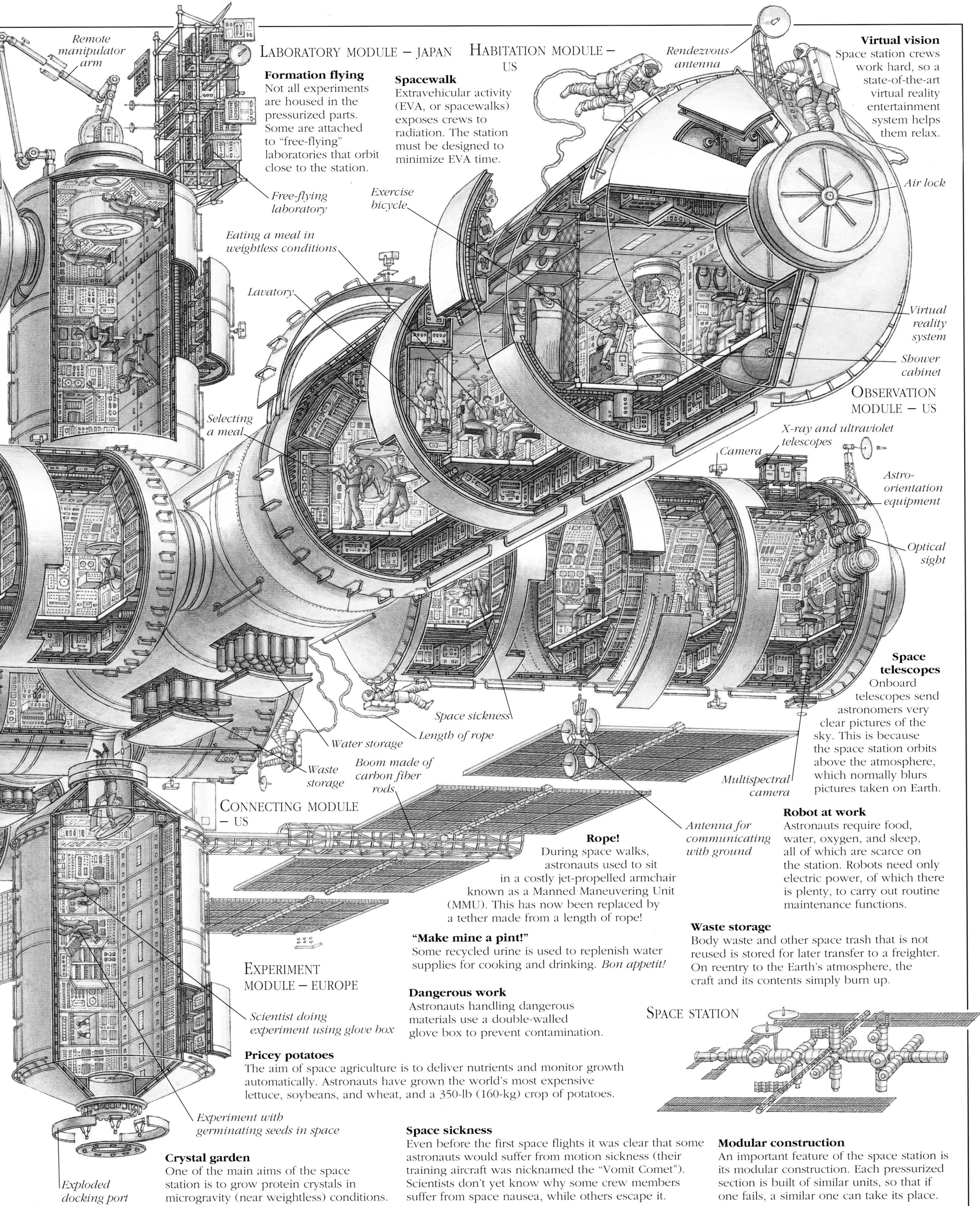

Formation flying
Not all experiments are housed in the pressurized parts. Some are attached to "free-flying" laboratories that orbit close to the station.

Spacewalk
Extravehicular activity (EVA, or spacewalks) exposes crews to radiation. The station must be designed to minimize EVA time.

Virtual vision
Space station crews work hard, so a state-of-the-art virtual reality entertainment system helps them relax.

Space telescopes
Onboard telescopes send astronomers very clear pictures of the sky. This is because the space station orbits above the atmosphere, which normally blurs pictures taken on Earth.

Rope!
During space walks, astronauts used to sit in a costly jet-propelled armchair known as a Manned Maneuvering Unit (MMU). This has now been replaced by a tether made from a length of rope!

Robot at work
Astronauts require food, water, oxygen, and sleep, all of which are scarce on the station. Robots need only electric power, of which there is plenty, to carry out routine maintenance functions.

"Make mine a pint!"
Some recycled urine is used to replenish water supplies for cooking and drinking. *Bon appetit!*

Waste storage
Body waste and other space trash that is not reused is stored for later transfer to a freighter. On reentry to the Earth's atmosphere, the craft and its contents simply burn up.

Dangerous work
Astronauts handling dangerous materials use a double-walled glove box to prevent contamination.

Pricey potatoes
The aim of space agriculture is to deliver nutrients and monitor growth automatically. Astronauts have grown the world's most expensive lettuce, soybeans, and wheat, and a 350-lb (160-kg) crop of potatoes.

Crystal garden
One of the main aims of the space station is to grow protein crystals in microgravity (near weightless) conditions.

Space sickness
Even before the first space flights it was clear that some astronauts would suffer from motion sickness (their training aircraft was nicknamed the "Vomit Comet"). Scientists don't yet know why some crew members suffer from space nausea, while others escape it.

Modular construction
An important feature of the space station is its modular construction. Each pressurized section is built of similar units, so that if one fails, a similar one can take its place.

Airport

If flying were as chaotic as driving, air travel would soon cease. The sky would be black with tiny aircraft. Midair collisions would be so common that newspapers wouldn't report them. At popular destinations, planes would wait for the lights to change, revving their engines and honking. Then they would race to see who was first down the runway. Of course, it wouldn't work. Flying is just too complex for each of us to own a plane. So at the airport we give up the privacy of our cars, we entrust our luggage to a stranger, and we share the journey with others. Perhaps if we did the same on roads, our world would be a safer, more peaceful, and more pleasant place.

Air traffic control
The job of air traffic control is to route aircraft safely to their destinations. To do this, controllers must make sure aircraft never come too close in flight, or when taking off and landing.

Keeping the skies safe
The approach control facility monitors aircraft beyond the circle handled by the visual control room.

Visual control tower

Lights out
Aircraft monitored by the approach control facility are tracked on radar screens. Older screens are dim, so controllers work in a darkened room.

Air traffic control

Approach control facility

Nice view
Aircraft within 5 miles (8 km) of the airport are under the control of staff in the visual control room.

Computer center

A souvenir, perhaps?
If you think the business of airports is transportation, you're wrong. Commercial operations, including gift shops, bars, and restaurants, bring in up to 60% of an airport's income.

Terminal 2nd Floor – departures (Green flow)

Bags of space
On average, every ten passengers on an international flight have 13 pieces of checked luggage.

Search me
X rays and searches stop would-be hijackers from carrying on guns. However, there is no quick way to screen all checked bags for bombs.

Check-in
All airlines have computerized check-in procedures. However, the procedure is simpler for domestic flights, and doesn't require the same degree of security.

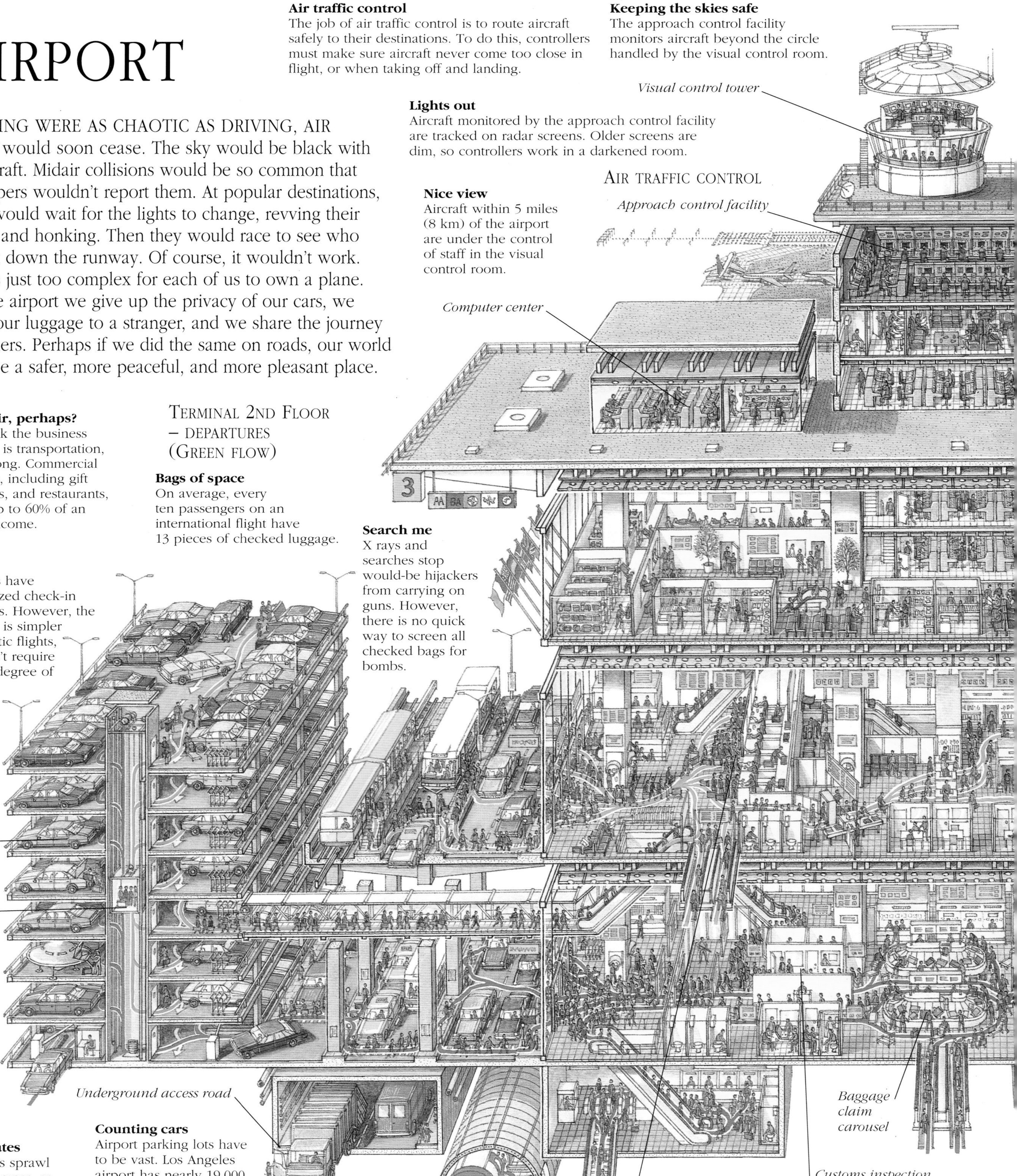

Multi-story parking lot

Elevator to floors

Parking lot road access

Underground access road

Distant gates
Big airports sprawl over huge areas, so "people movers" are often needed for passengers transferring between flights. When Dallas-Fort Worth airport is complete, its most distant gates will be 4.5 miles (7 km) apart.

Counting cars
Airport parking lots have to be vast. Los Angeles airport has nearly 19,000 spaces – if all the cars left at once, they'd form a line nearly 60 miles (100 km) long.

Subway train

Parking ticket or phone number?
As anybody who has driven to an airport knows, the parking lot is always expensive. It provides up to a sixth of an airport's profit.

Passenger check-in

Baggage claim carousel

Customs inspection area

Not-so-rapid access
Rapid transit systems linking major airports with cities are popular with passengers: half of the passengers to England's Gatwick airport take the train.

Take the train
The belts carrying luggage are like a miniature railroad complex within the terminal: Frankfurt airport in Germany has 25 miles (40 km) of track.

Automatic chaos
Computers help plan flights, but controllers still retain a manual system. If the computers break down, they write details of each flight on strips of cardboard, and shuffle them into a sequence of landing and takeoff slots.

Runway
The paved strip on which aircraft land is called the runway. Its length depends on the type of aircraft using it and the altitude and climate of the airport.

Runway markings
Pilots use painted markings to judge their approach by day in good weather. A standard pattern of stripes marks the center-line and boundaries and helps pilots judge their altitude.

Green for go
A green bar marks the threshold – the start of the runway.

Approach lights
A standard pattern of lights visible only from the air points the way to the runway.

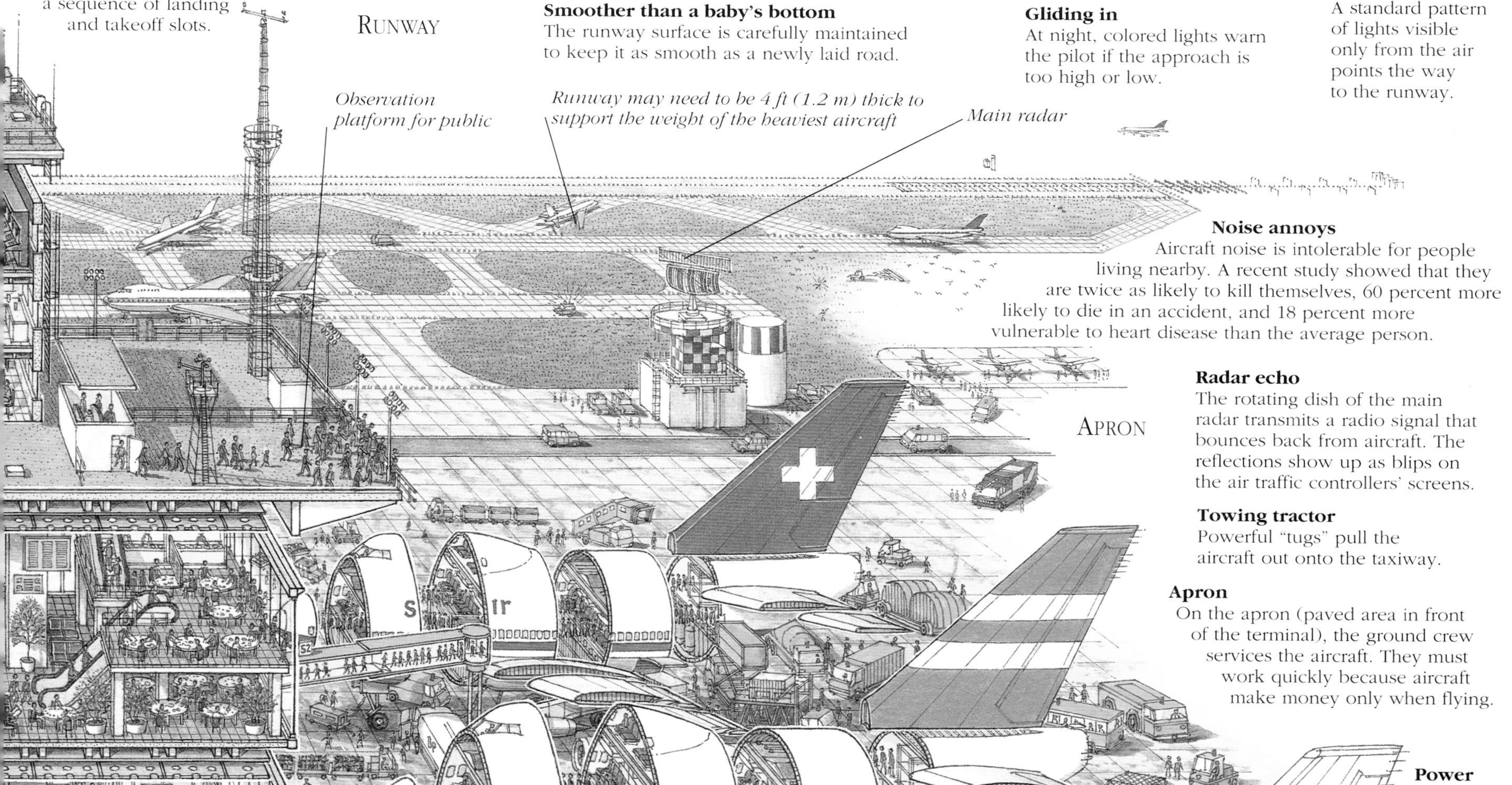

Smoother than a baby's bottom
The runway surface is carefully maintained to keep it as smooth as a newly laid road.

Gliding in
At night, colored lights warn the pilot if the approach is too high or low.

Noise annoys
Aircraft noise is intolerable for people living nearby. A recent study showed that they are twice as likely to kill themselves, 60 percent more likely to die in an accident, and 18 percent more vulnerable to heart disease than the average person.

Radar echo
The rotating dish of the main radar transmits a radio signal that bounces back from aircraft. The reflections show up as blips on the air traffic controllers' screens.

Towing tractor
Powerful "tugs" pull the aircraft out onto the taxiway.

Apron
On the apron (paved area in front of the terminal), the ground crew services the aircraft. They must work quickly because aircraft make money only when flying.

Power hungry
Aircraft engines generate electricity, but the supply stops when the engines do. A portable generator then supplies power.

Sorting it out
Luggage on the conveyor belt is usually hand sorted, but a few airports have an automatic system. This reads flight numbers from bar codes on tags. The system pushes the bags off the main belt into "branch lines" that lead to the right aircraft.

You're welcome!
The most welcoming airport in the world is on the island of Curaçao. Two-thirds of the people visiting the airport go there to welcome their friends or family, or to say "farewell." The least welcoming airport is Paris – only 7 percent of visitors are greeters and senders.

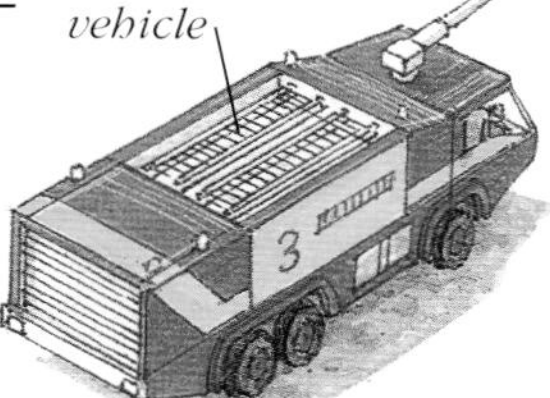

Fill 'er up!
Many airports have hydrants that pump aviation fuel from huge reservoirs. England's Heathrow Airport has a store of 15.6 million gallons (60 million liters).

"Please, not the hovercraft!"
Firefighters are ready for everything. One major European airport is near a sewage treatment plant, and rescue teams are equipped with an inflatable boat and hovercraft in case a plane overshoots the runway and lands in a sludge pond.

Line of luggage
A typical 747 carries 800 pieces of checked luggage. Laid side-by-side, the suitcases would stretch six times the length of the aircraft.

Get a move on!
Luggage from an arriving 747 can be waiting on the baggage carousels within 12 minutes, but some airlines take three times as long.

Industrial revolution

Canals
Canals were the highways of the early 19th century. A single horse could pull a heavy canal boat. On even a good road, the same horse could move only 1/100th of the load.

Coal mine
Coal fueled the industrial revolution. At some mines pit-head winding gear winched tubs of coal from the seams deep underground. However, in most pits, children climbed ladders to carry the coal out of the mine.

Railroads
The first railroad trains hauled freight cars of coal rather than passenger cars.

Fire

Big burns
Overcrowding has made large cities a terrible fire risk throughout history. Fires destroyed Dresden in 1491, London in 1666, Chicago in 1871, and San Francisco in 1906.

London's burning!
A huge fire broke out in London in 1666, but it did not appear serious at first. When the fire started, the Lord Mayor of London dismissed the flames with the words "Pish, a woman might piss it out!" and went back to bed.

Flaming feathers
Diary-writer Samuel Pepys (1633-1703) watched the Great Fire of London and felt sorry for the pigeons, saying they "hovered about the windows ... till they burned their wings and fell down."

Fire fighting
17th-century firefighters had little chance against a big blaze. They hauled water in tubs and pumped it onto the flames by hand.

Plague

"This is the end of the world..."
As large families crowded into small city houses, disease followed close behind. The biggest killer was bubonic plague – a disease spread by the fleas that lived on rats. An epidemic in the 14th century killed between 1/4 and 1/2 of Europe's population.

Let's get out of here!
Rich people fled when there was a plague. The poor had nowhere to go, nor could they afford to travel, so they stayed – and died.

Sealing a house
To prevent the spread of plague, city officials sealed the houses of victims so that the affected family could not come out and infect others.

Plague pits
As more and more people died, the bodies were tipped into huge "plague pits," or mass graves.

Vikings

Vikings
Viking peoples advanced from Scandinavia to Europe between AD 800 and 1100, bringing with them traditions and skills.

Longboat
Skilled seamanship made possible Viking expansion. In their broad open boats, the Vikings even crossed the Atlantic Ocean.

Viking houses
Viking houses looked like upturned boats, and in ancient times they may even have been roofed with an old, leaky boat.

City cross-section

Time machines really do exist. Archaeologists – scientists who study how our ancestors lived – use them to travel back into the past. They call their time machines "spades." With the aid of spades, trowels, and even soft brushes, archaeologists carefully remove layer upon layer of earth. Digging deeper and deeper, they travel back in time, for they know that, on any one spot, the objects close to the surface are more recent; those buried deepest are the oldest. As archaeologists sift through what remains of past people, they piece together pictures of vanished ages. Far below the city streets lies solid rock. Now paleontologists take up the journey. They study fossils – stone casts of the animals and plants from the past. From fossils they can judge what life was like before humans arrived. The oldest rocks of all contain no fossils. Formed thousands of millions of years ago, before life began, they are a reminder that our Earth was once a ball of dust spinning in cold black space.

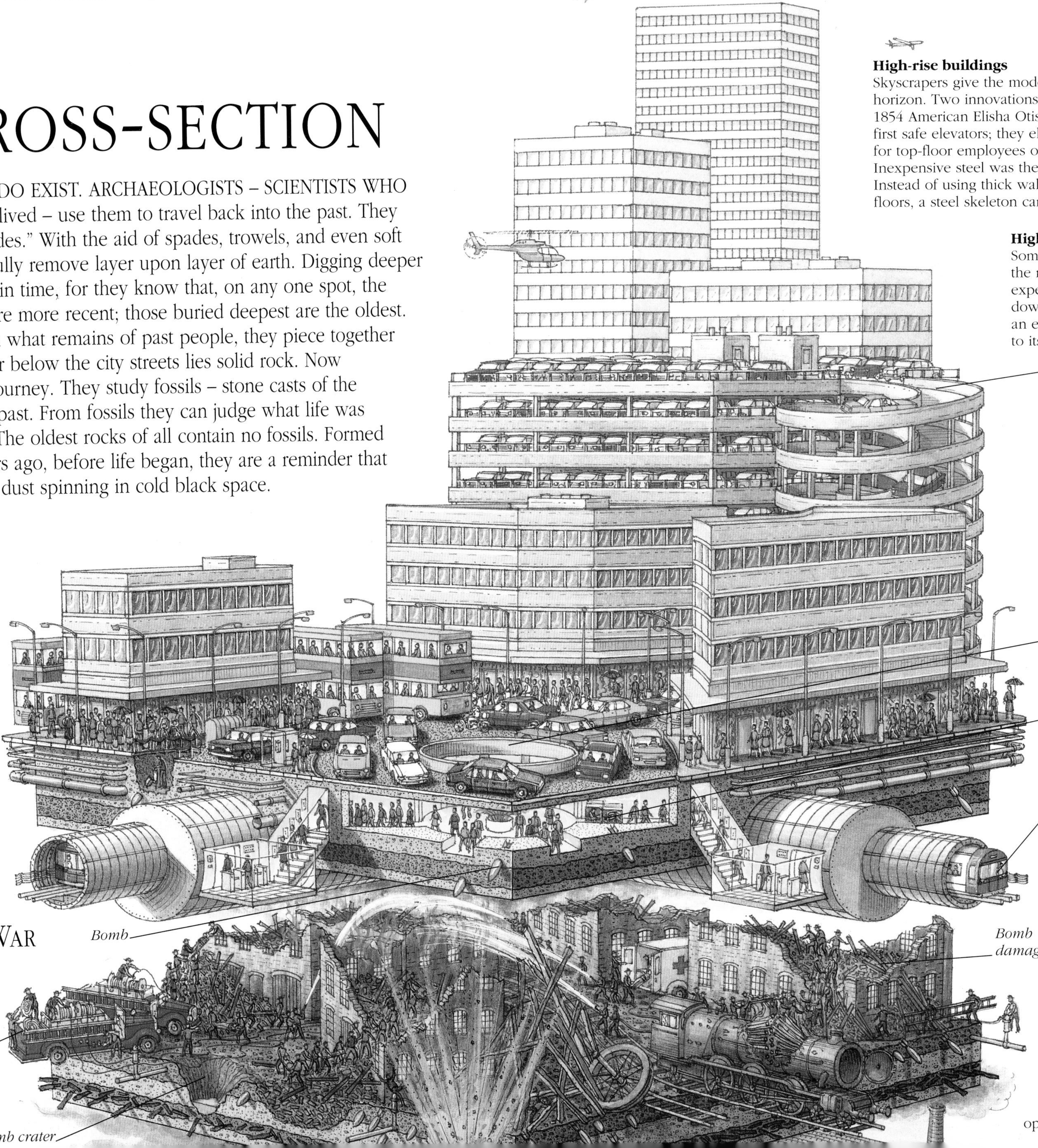

Today

Underneath the sidewalk
Underneath pedestrians' feet run the arteries of the city. Gas and water flow along pipes, and electric cables supply power. Sewers take waste away from buildings, and communications cables transport data and phone calls.

Dangerous digging
Workers on any city building site have to thread their way gingerly through mazes of underground pipes and cables. In some European cities, unexploded bombs create more hazards.

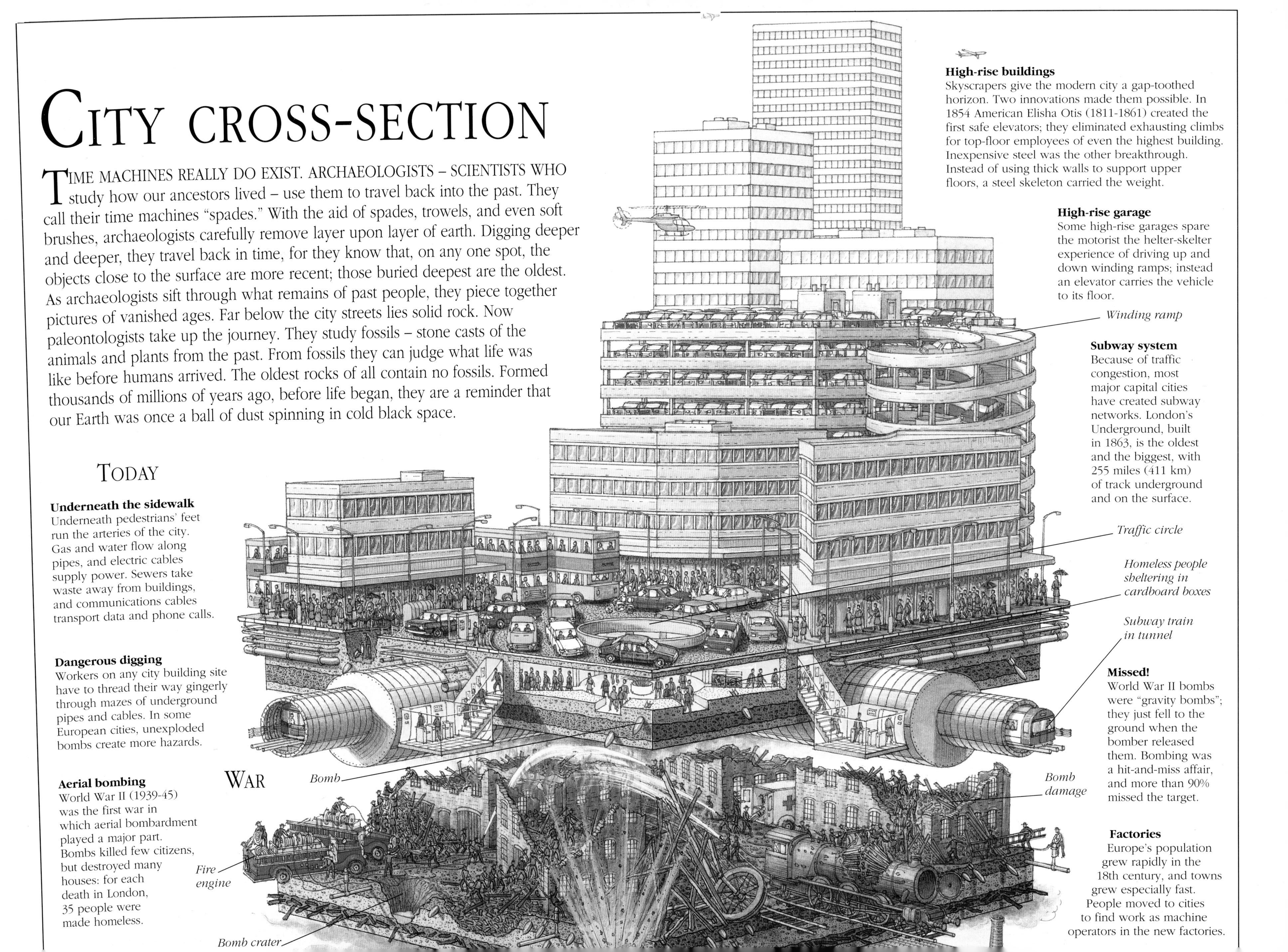

High-rise buildings
Skyscrapers give the modern city a gap-toothed horizon. Two innovations made them possible. In 1854 American Elisha Otis (1811-1861) created the first safe elevators; they eliminated exhausting climbs for top-floor employees of even the highest building. Inexpensive steel was the other breakthrough. Instead of using thick walls to support upper floors, a steel skeleton carried the weight.

High-rise garage
Some high-rise garages spare the motorist the helter-skelter experience of driving up and down winding ramps; instead an elevator carries the vehicle to its floor.

Winding ramp

Subway system
Because of traffic congestion, most major capital cities have created subway networks. London's Underground, built in 1863, is the oldest and the biggest, with 255 miles (411 km) of track underground and on the surface.

Traffic circle

Homeless people sheltering in cardboard boxes

Subway train in tunnel

War

Aerial bombing
World War II (1939-45) was the first war in which aerial bombardment played a major part. Bombs killed few citizens, but destroyed many houses: for each death in London, 35 people were made homeless.

Bomb

Fire engine

Bomb crater

Bomb damage

Missed!
World War II bombs were "gravity bombs"; they just fell to the ground when the bomber released them. Bombing was a hit-and-miss affair, and more than 90% missed the target.

Factories
Europe's population grew rapidly in the 18th century, and towns grew especially fast. People moved to cities to find work as machine operators in the new factories.

WINDMILL

Common sails
The simplest type of sails had a covering of canvas to catch the wind. To stop the sails from turning dangerously fast in high winds, the miller reefed (tied back) part of the canvas to reduce the sail area.

Reefing names
Millers gave names to the different degrees of reefing. This sail is set to "sword point." Tying in more canvas set the sails to "dagger point." Spreading the canvas set them to "first reef."

Haul away!
The miller raised sacks of gr a winch powered by the sai boy tied on sacks at ground jerked on the control rope t "haul away" to the miller ab

Wind shaft
The sails rotated on the end wind shaft. A small trapdoor wind shaft allowed the mille the sails for maintenance.

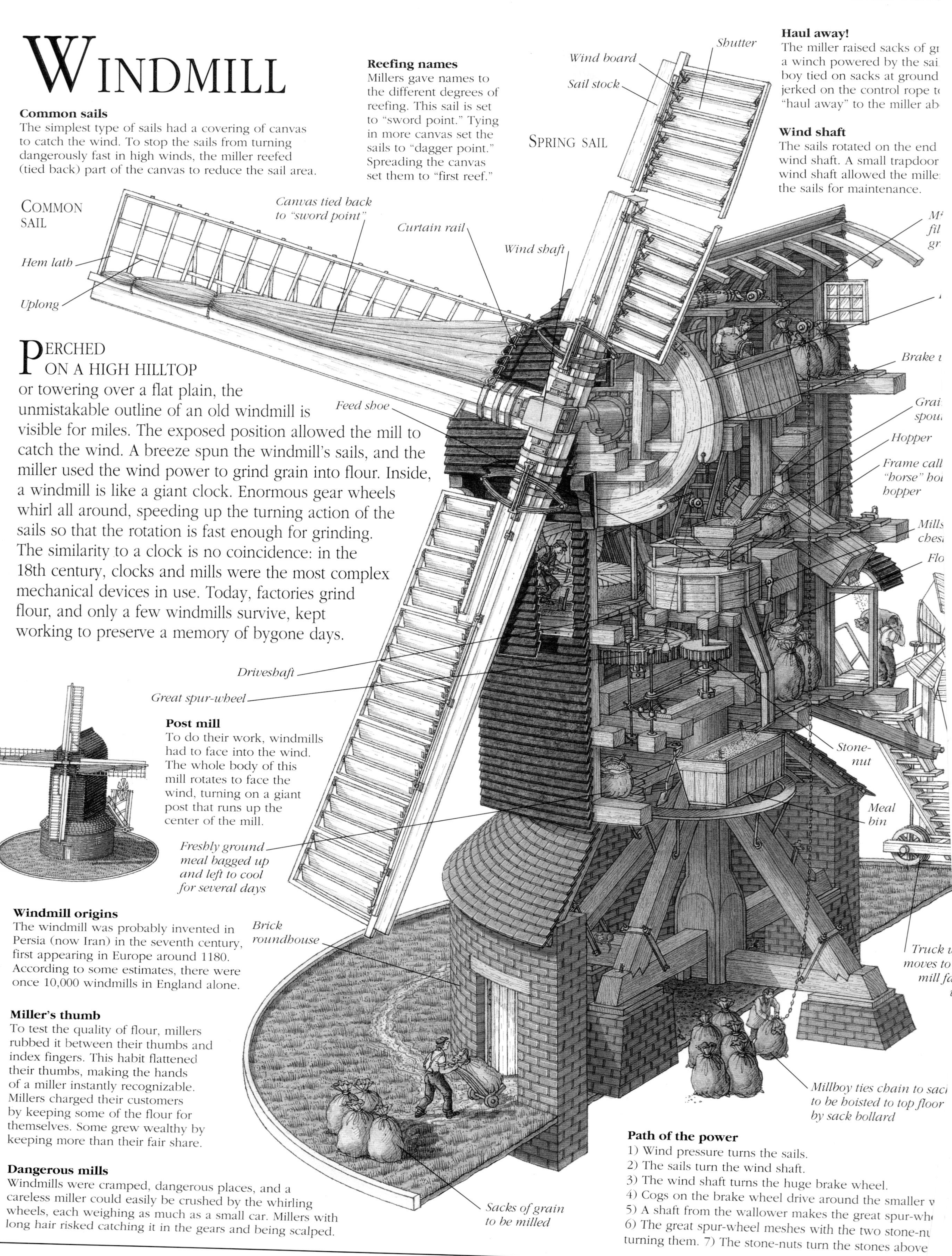

PERCHED ON A HIGH HILLTOP or towering over a flat plain, the unmistakable outline of an old windmill is visible for miles. The exposed position allowed the mill to catch the wind. A breeze spun the windmill's sails, and the miller used the wind power to grind grain into flour. Inside, a windmill is like a giant clock. Enormous gear wheels whirl all around, speeding up the turning action of the sails so that the rotation is fast enough for grinding. The similarity to a clock is no coincidence: in the 18th century, clocks and mills were the most complex mechanical devices in use. Today, factories grind flour, and only a few windmills survive, kept working to preserve a memory of bygone days.

Post mill
To do their work, windmills had to face into the wind. The whole body of this mill rotates to face the wind, turning on a giant post that runs up the center of the mill.

Windmill origins
The windmill was probably invented in Persia (now Iran) in the seventh century, first appearing in Europe around 1180. According to some estimates, there were once 10,000 windmills in England alone.

Miller's thumb
To test the quality of flour, millers rubbed it between their thumbs and index fingers. This habit flattened their thumbs, making the hands of a miller instantly recognizable. Millers charged their customers by keeping some of the flour for themselves. Some grew wealthy by keeping more than their fair share.

Dangerous mills
Windmills were cramped, dangerous places, and a careless miller could easily be crushed by the whirling wheels, each weighing as much as a small car. Millers with long hair risked catching it in the gears and being scalped.

Path of the power
1) Wind pressure turns the sails.
2) The sails turn the wind shaft.
3) The wind shaft turns the huge brake wheel.
4) Cogs on the brake wheel drive around the smaller w
5) A shaft from the wallower makes the great spur-wh
6) The great spur-wheel meshes with the two stone-nu turning them. 7) The stone-nuts turn the stones above

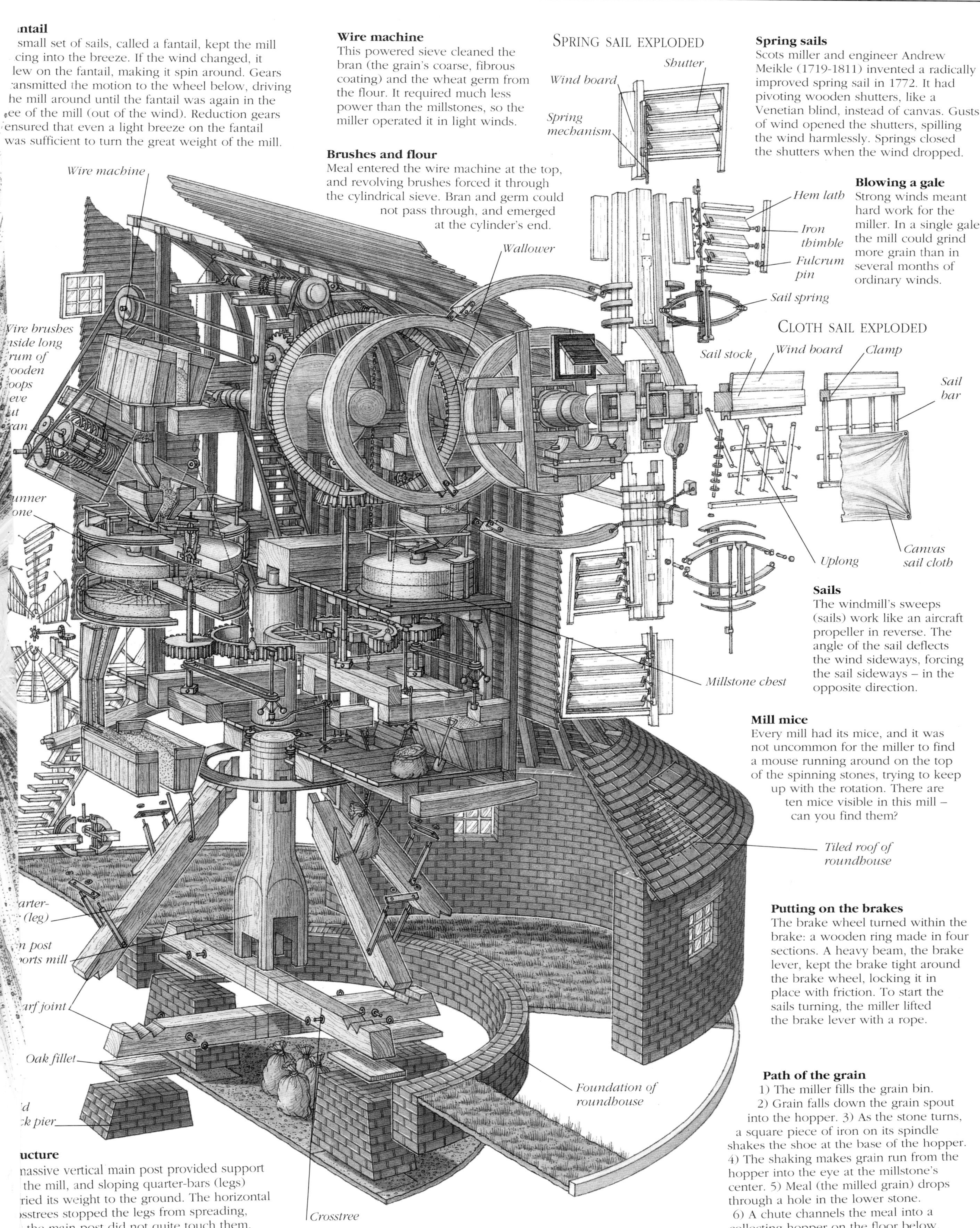

ıntail
small set of sails, called a fantail, kept the mill cing into the breeze. If the wind changed, it lew on the fantail, making it spin around. Gears ansmitted the motion to the wheel below, driving he mill around until the fantail was again in the ee of the mill (out of the wind). Reduction gears ensured that even a light breeze on the fantail was sufficient to turn the great weight of the mill.

Wire machine
This powered sieve cleaned the bran (the grain's coarse, fibrous coating) and the wheat germ from the flour. It required much less power than the millstones, so the miller operated it in light winds.

Brushes and flour
Meal entered the wire machine at the top, and revolving brushes forced it through the cylindrical sieve. Bran and germ could not pass through, and emerged at the cylinder's end.

Spring sails
Scots miller and engineer Andrew Meikle (1719-1811) invented a radically improved spring sail in 1772. It had pivoting wooden shutters, like a Venetian blind, instead of canvas. Gusts of wind opened the shutters, spilling the wind harmlessly. Springs closed the shutters when the wind dropped.

Blowing a gale
Strong winds meant hard work for the miller. In a single gale the mill could grind more grain than in several months of ordinary winds.

Sails
The windmill's sweeps (sails) work like an aircraft propeller in reverse. The angle of the sail deflects the wind sideways, forcing the sail sideways – in the opposite direction.

Mill mice
Every mill had its mice, and it was not uncommon for the miller to find a mouse running around on the top of the spinning stones, trying to keep up with the rotation. There are ten mice visible in this mill – can you find them?

Putting on the brakes
The brake wheel turned within the brake: a wooden ring made in four sections. A heavy beam, the brake lever, kept the brake tight around the brake wheel, locking it in place with friction. To start the sails turning, the miller lifted the brake lever with a rope.

Path of the grain
1) The miller fills the grain bin. 2) Grain falls down the grain spout into the hopper. 3) As the stone turns, a square piece of iron on its spindle shakes the shoe at the base of the hopper. 4) The shaking makes grain run from the hopper into the eye at the millstone's center. 5) Meal (the milled grain) drops through a hole in the lower stone. 6) A chute channels the meal into a collecting hopper on the floor below.

ucture
nassive vertical main post provided support the mill, and sloping quarter-bars (legs) ried its weight to the ground. The horizontal osstrees stopped the legs from spreading, the main post did not quite touch them.

Early seas

Teeming life
390 to 340 million years ago, during the Devonian period, teeming life in the warm seas began to evolve into the very first land animals.

First amphibians
Animals first struggled out of the water using their fins as primitive legs and gasping for breath with their inefficient lungs. Millions of years passed before they became fully adapted to life on land.

Simple molecules of life
Scientists can only speculate at how life began on Earth. However, the first living things on our planet were bacteria.

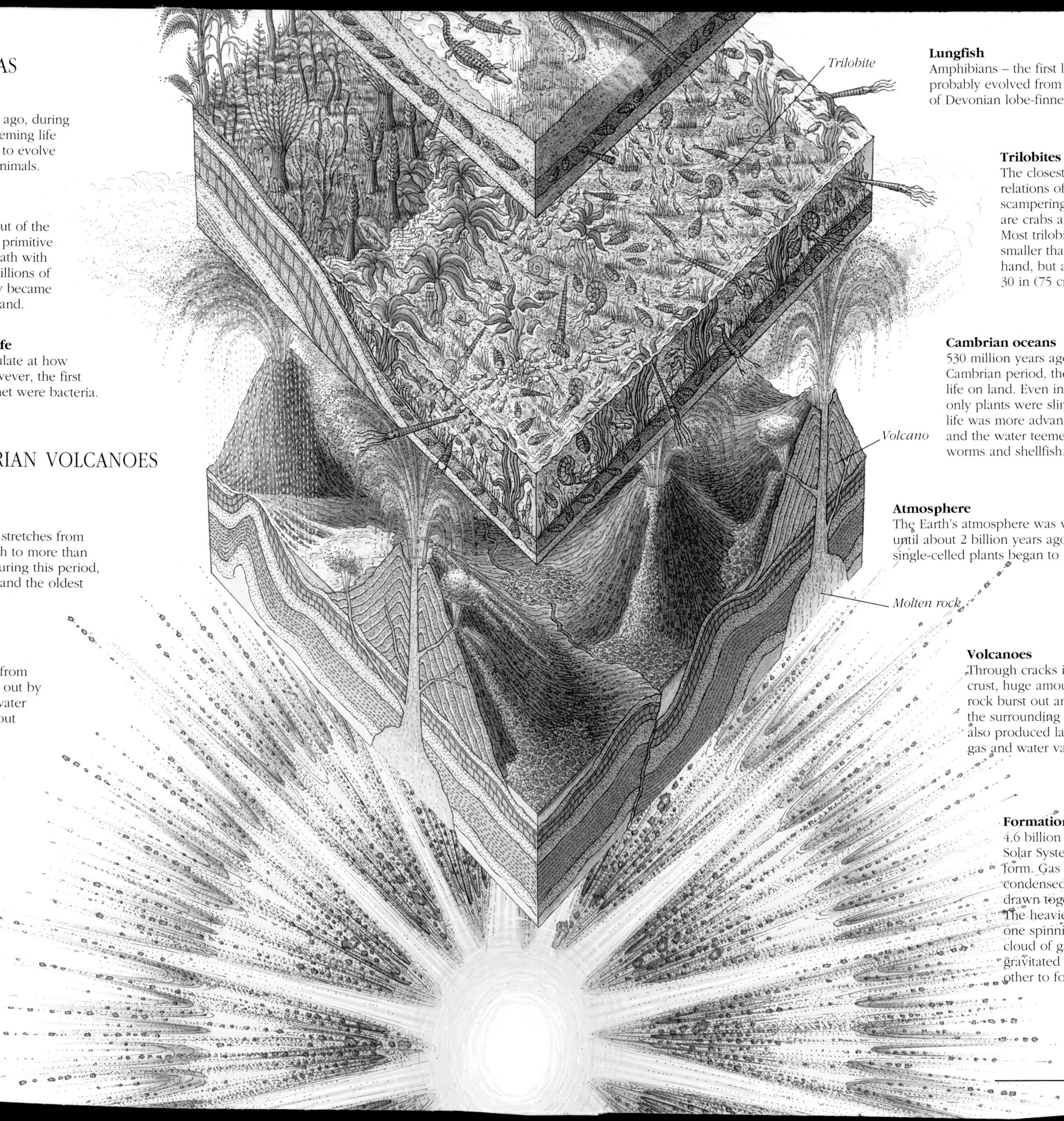

Lungfish
Amphibians – the first land animals – probably evolved from three groups of Devonian lobe-finned fishes.

Trilobites
The closest living relations of these scampering sea creatures are crabs and lobsters. Most trilobites were smaller than an adult's hand, but a few grew to 30 in (75 cm) in length.

Cambrian oceans
530 million years ago, during the Cambrian period, there was no plant life on land. Even in the oceans, the only plants were slimy algae. Animal life was more advanced, however, and the water teemed with primitive worms and shellfish.

Trilobite

Volcano

Precambrian volcanoes

The cooling Earth
The Precambrian period stretches from the formation of the Earth to more than 570 million years ago. During this period, the Earth's crust cooled, and the oldest rocks were forming.

Water
Seas and oceans formed from the water vapor pumped out by volcanoes. Pools of rainwater began to collect from about 3.8 billion years ago.

Atmosphere
The Earth's atmosphere was very thin until about 2 billion years ago, when the first single-celled plants began to produce oxygen.

Volcanoes
Through cracks in the Earth's crust, huge amounts of molten rock burst out and flowed over the surrounding land. Volcanoes also produced large clouds of gas and water vapor.

Molten rock

Big bang

Big Bang
The Universe may have begun as a single tiny particle with infinite mass. 10 billion years ago it flew apart in a massive explosion that we call the Big Bang.

Formation of the Earth
4.6 billion years ago the Solar System began to form. Gas and dust condensed in space, drawn together by gravity. The heaviest particles in one spinning, eddying cloud of gas and matter gravitated toward each other to form the Earth.

Romans

Conquerors
Spreading outward from Italy, Roman warriors conquered almost all of southern and western Europe by AD 230.

Fighting queen
The Roman rulers of eastern England whipped Celtic Queen Boudicca (?–AD 62) when her husband died. They then stole her land. Boudicca fought back by leading a rebellion in which her fighters slaughtered thousands of their hated Roman masters.

Orderly advance
Roman troops were highly disciplined fighters. When under attack, the troops could raise their long shields to make a continuous protective roof over their heads.

Pay day tomorrow
Roman soldiers had a remarkably modern salary system. Deductions from their pay even included a tombstone club, which paid for a funeral in installments.

Early people

Emerging structures
4-5,000 years ago, the peoples of western Europe began to farm and domesticate animals – rather than gathering fruits, roots, and seeds and hunting. This more settled way of life gave them time to develop social structures and religions.

Beaker people
About 4,000 BC communities of Beaker people spread across Europe. Archaeologists named them for the characteristic beaker pottery they made.

Mammoth
The elephantlike woolly mammoth stalked Europe during the Pleistocene epoch, which began some 2.5 million years ago.

Megaliths
Neolithic people all over Europe built megaliths – upright stones, often arranged in circles and lines. Most scholars agree that they must have been used as part of religious celebrations.

Heavy haulage
Some of the most impressive stone circles in Europe are amazing feats of engineering. Their builders moved huge stones up to 200 miles (320 km).

Pitfall trap
Teams of hunters dug pits, concealing them with brushwood. Panicking the mammoths drove them across the pits and they fell in.

Vicious weapons
Ancient hunters may have used stone-tipped javelins to hunt mammoths.

Dinosaurs

Jurassic forest
Far beneath the streets of the modern city lie the remains of creatures that lived 150 million years ago. Most remarkable were the dinosaurs. Some were huge vegetarians that browsed on trees. Many dinosaurs, though, were ferocious meat eaters.

Apatosaurus
This huge plant eater weighed as much as four modern-day adult African elephants. It wandered the Jurassic forests of what is now the western United States.

Allosaurus
Though only a third of its size, *Allosaurus* made a meal of *Apatosaurus*. Paleontologists are not sure, though, how *Allosaurus* fed. Some think that this dinosaur was a scavenger, feeding only on the remains of already dead animals.

ANTARCTIC BASE

IN THE MOST ISOLATED SPOT ON EARTH, THE SOUTH POLE, THE SUN NEVER SETS IN the summer, and winter is one long night. The climate is so hostile that no plant or animal can survive for long in the open. The South Pole lies at the heart of Antarctica, a vast ice-covered continent. Antarctica was largely unexplored until seventy years ago. Today, scientists line up to work in this cold and remote location. Why? Because astronomers can peer at the stars for months on end, without daytime interrupting their work; biologists gain a super-clean laboratory in the purest air on Earth. Even the ice cap itself is a remarkable archive for study. Its layers of compressed snow comprise a 150,000-year-old record of the Earth's climate.

Polar night
"Night" begins in March with a sunset that lasts a month. "Morning" comes six months later. In the winter sky the stars circle continuously, without rising or setting as they do elsewhere on Earth.

Freewheeling
Tracked vehicles are the main form of transportation at the pole. Coastal bases also use huge wheeled trucks called Deltas, plus motorcycles and snowmobiles.

Explorer's memorial
This drawing is based on the United States Amundsen-Scott South Pole Station. Roald Amundsen (1872-1928) planted the Norwegian flag at the South Pole on December 14, 1911. The expedition led by British explorer Robert Scott (1869-1912) reached the pole a month later, but his team died on the return journey.

Underneath the arches
Steel arches provide much of the indoor space at the base. Their shape has been chosen to shed as much snow as possible.

The ozone hole
One of the most important jobs of the polar base is to monitor the Earth's ozone layer. Each year a hole in the ozone layer opens above Antarctica, and scientists at the base use their instruments to judge how much of a threat this poses.

Ice on the move
The ice sheet on which the polar station rests is moving slowly. The base moves with it.

Don't throw water
The low temperatures at the base freeze all water vapor, drying wood buildings until they are like tinder. Fire is always a hazard.

Work in progress
Maintaining the base demands constant work. Many of the structures are nearing the end of their useful life, and the whole base will need replacing in a few years.

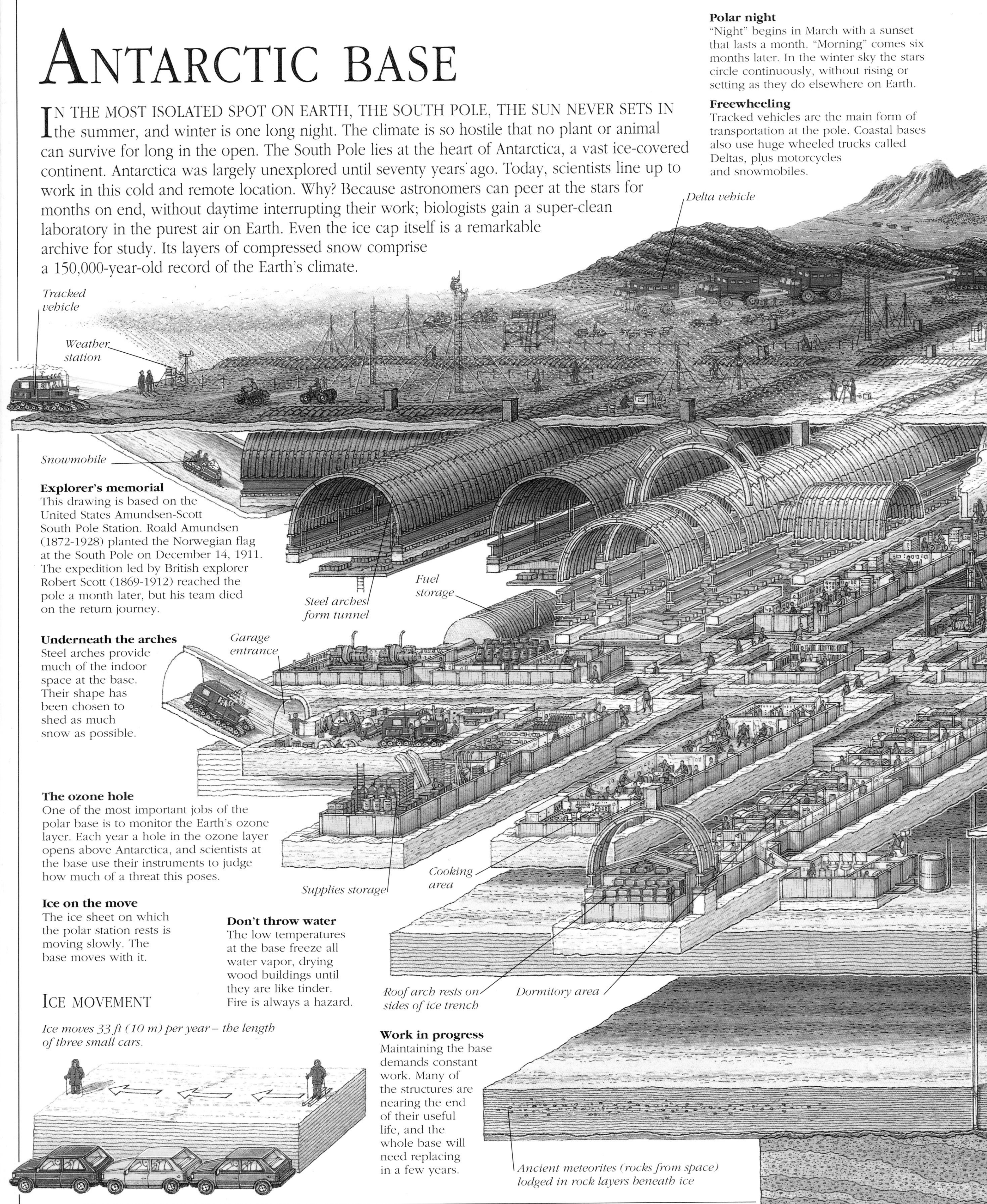

Polar day
At the pole, day and night don't take turns as they do on the rest of the planet. In "daytime" – which lasts the whole summer – the sun never sets: instead it just circles in the sky.

Airdrop
Supplies for the South Pole base are ferried in by air. In summer, aircraft can land on a prepared strip. Winter supplies drop by parachute.

Doomed dome
The most visible building at the base is the 165-ft (50-m) wide geodesic dome. However, the aluminum dome is gradually being covered by snow and ice, and this will eventually crush it.

Clear skies mean better views
Astronomers like using their telescopes at the pole because the sky is exceptionally clear. The biting frost freezes all the water vapor that normally blurs observations out of the atmosphere.

Eco-tourism
Sight-seeing flights do little harm to the fragile polar environment, but the impact of the 4,000 wealthy tourists who visit each year is controversial. Their 8,000 boots can damage coastal lichens that take decades to grow.

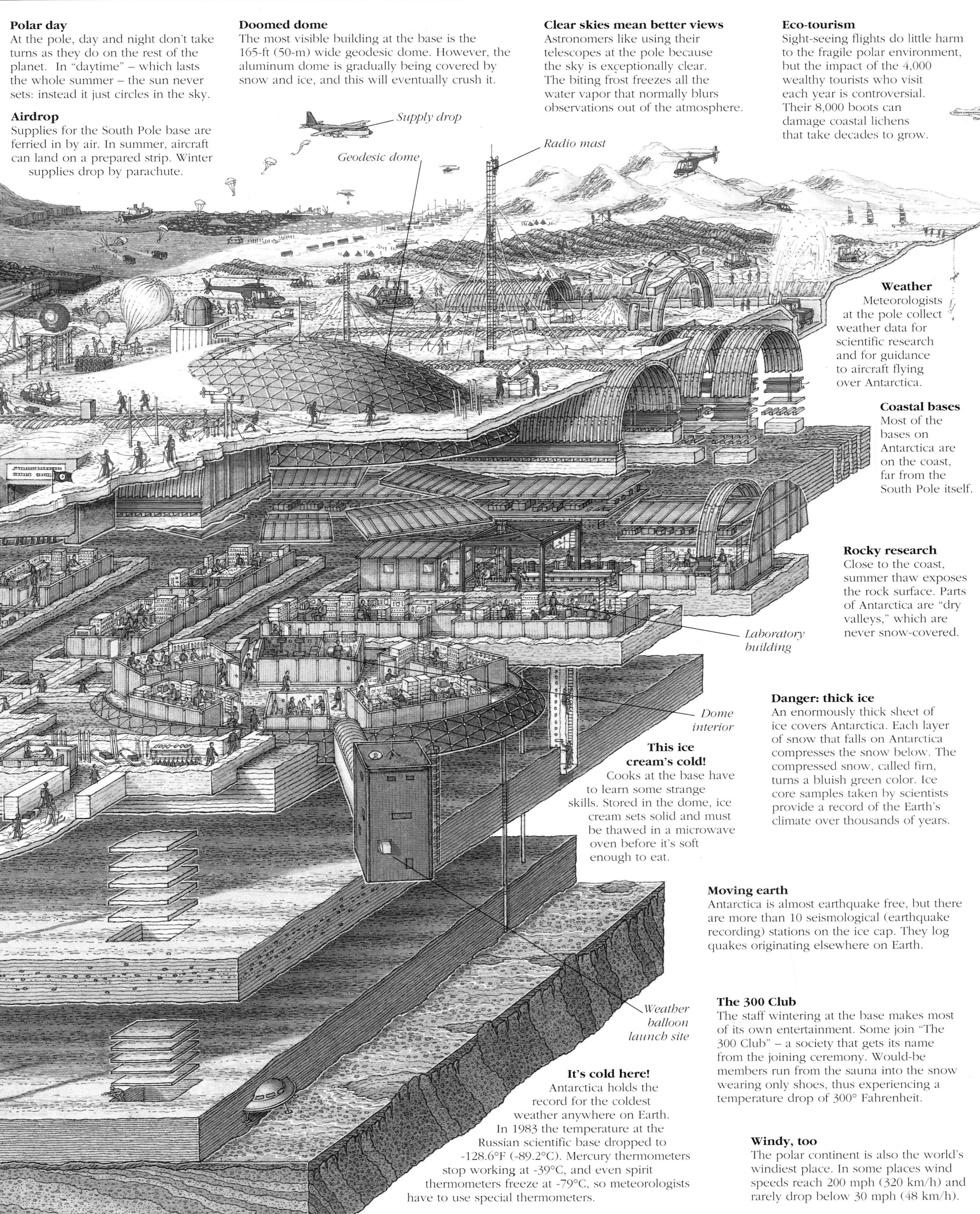

Weather
Meteorologists at the pole collect weather data for scientific research and for guidance to aircraft flying over Antarctica.

Coastal bases
Most of the bases on Antarctica are on the coast, far from the South Pole itself.

Rocky research
Close to the coast, summer thaw exposes the rock surface. Parts of Antarctica are "dry valleys," which are never snow-covered.

Danger: thick ice
An enormously thick sheet of ice covers Antarctica. Each layer of snow that falls on Antarctica compresses the snow below. The compressed snow, called firn, turns a bluish green color. Ice core samples taken by scientists provide a record of the Earth's climate over thousands of years.

This ice cream's cold!
Cooks at the base have to learn some strange skills. Stored in the dome, ice cream sets solid and must be thawed in a microwave oven before it's soft enough to eat.

Moving earth
Antarctica is almost earthquake free, but there are more than 10 seismological (earthquake recording) stations on the ice cap. They log quakes originating elsewhere on Earth.

The 300 Club
The staff wintering at the base makes most of its own entertainment. Some join "The 300 Club" – a society that gets its name from the joining ceremony. Would-be members run from the sauna into the snow wearing only shoes, thus experiencing a temperature drop of 300° Fahrenheit.

It's cold here!
Antarctica holds the record for the coldest weather anywhere on Earth. In 1983 the temperature at the Russian scientific base dropped to -128.6°F (-89.2°C). Mercury thermometers stop working at -39°C, and even spirit thermometers freeze at -79°C, so meteorologists have to use special thermometers.

Windy, too
The polar continent is also the world's windiest place. In some places wind speeds reach 200 mph (320 km/h) and rarely drop below 30 mph (48 km/h).

MOVIE STUDIO

IN THE WORLD OF MOVIES, NOTHING IS QUITE WHAT IT SEEMS. BUILDINGS that seem to tower on-screen are just tiny models. The giant squid dragging a ship beneath the waves is a latex puppet. A space station turns out to be a large sheet of canvas. Making these illusions look convincing requires the skills of a huge team of people, including set designers, background painters, and modelmakers. These specialists craft the film in a studio complex. Team members may work for months on the tiny details of a set – only to see it flash by on the screen in a few seconds.

Studio city
The soundstages, where filming takes place, are like huge sheds. They are carefully soundproofed, so that the sensitive microphones don't pick up the noise from traffic and aircraft outside. Although this drawing shows filming taking place on three sets all at once, normally only one of them would be in use at any time.

The backroom boys (and girls)
The soundstage is vast and obvious, but it's only a small fraction of the whole. Clustered around the cavernous set there's a maze of small workshops. Inside them an army of technicians and support staff toil. Without their support the production would sink.

Animatronics
Modelmakers mold latex foam into elaborate monster masks and creatures such as a small squid. Tiny motors hidden inside the rubber skin control the movement of the limbs and facial expressions.

Distant illusion
Carpenters build a quarter-scale replica of a Buddha statue. When it appears in the background it will look like the full-size statue, four times farther away. Without such perspective tricks, the studio would need to be four times as big.

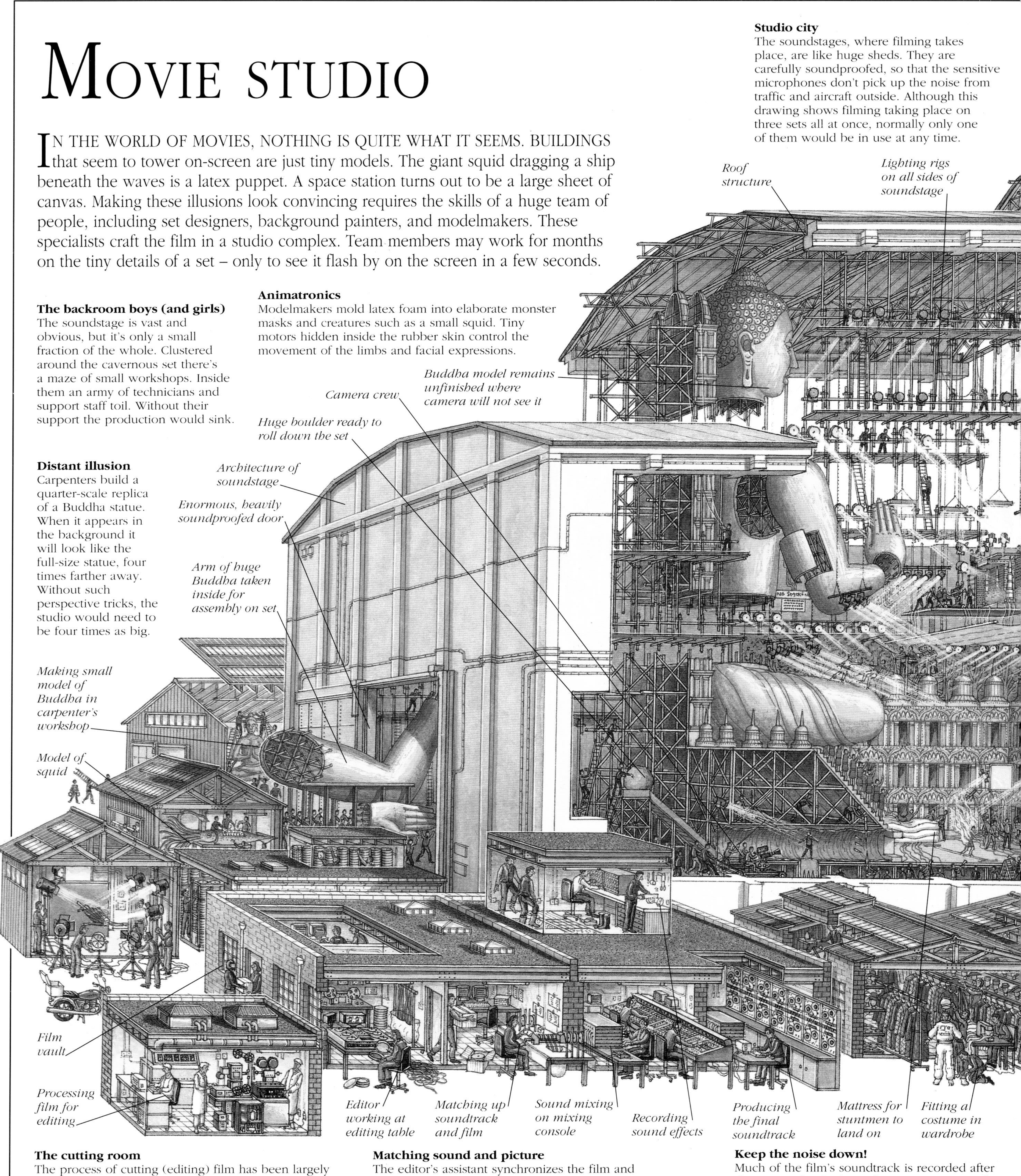

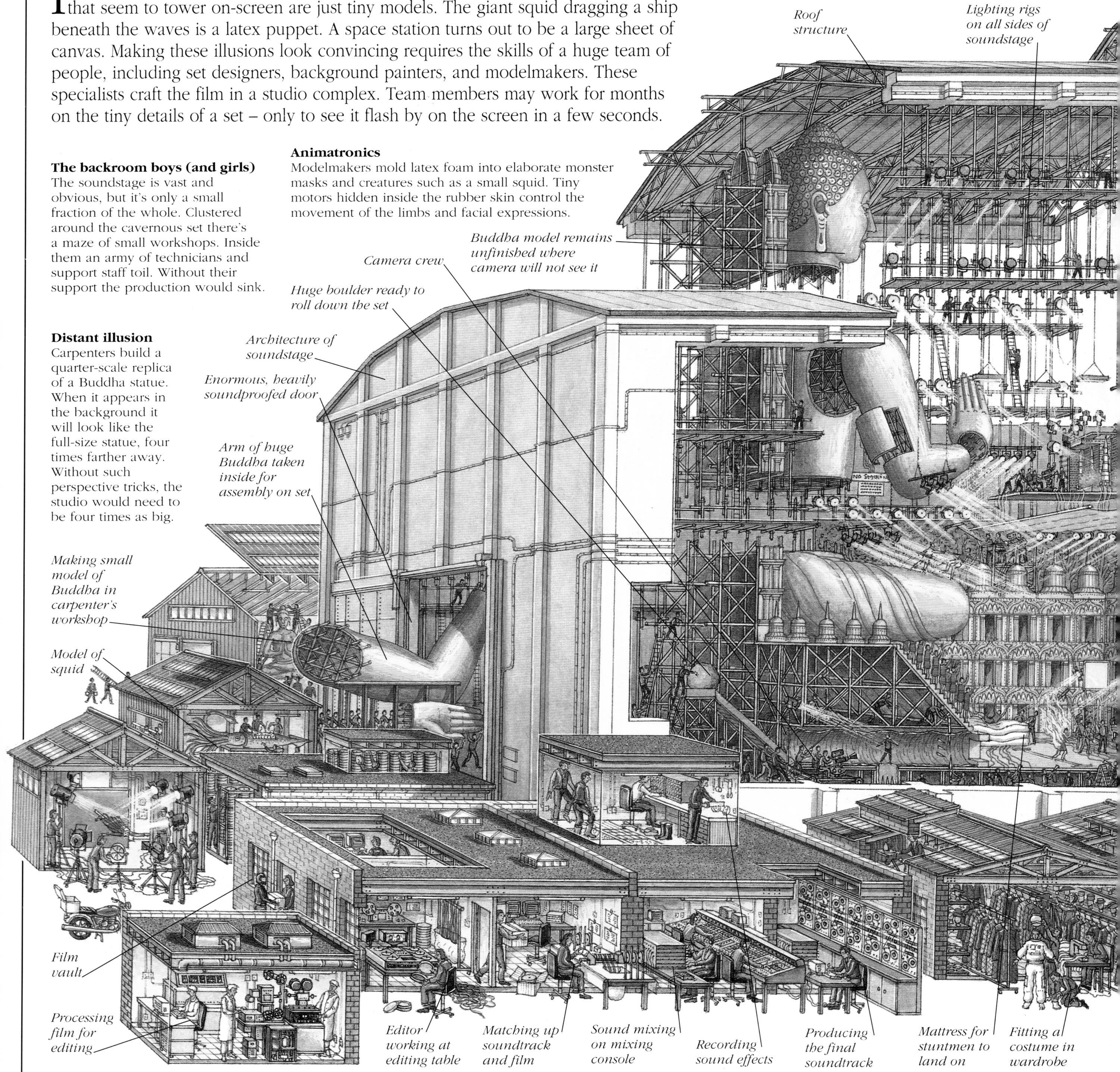

The cutting room
The process of cutting (editing) film has been largely unchanged by technology. Most editors still cut and join a cutting print (positive copy) of film shot in the camera. This will be used as a guide for cutting the negative from which the final prints are made.

Matching sound and picture
The editor's assistant synchronizes the film and soundtrack, which is recorded on magnetic tape with sprocket holes just like film. The assistant checks that the "clap" sound at the start of each scene lines up with the film frame showing the clapperboard closing.

Keep the noise down!
Much of the film's soundtrack is recorded after shooting is finished. For footsteps, actors walk around on simulated patches of "gravel path" or "fall leaves." The final sound of the film may consist of thirty or forty tracks mixed together.

Big building
Soundstages can be vast. The biggest, at Pinewood Studios in England, could hold more than 570 double-decker buses.

Dry ice inside rocket for blast-off effect

Huge rocket model for sci-fi film

Speedy spray guns
Painted backdrops are often used because they are so cheap and quick to produce. Two painters might take just a few days to conjure up a background the size of a tennis court.

Studio city
The biggest studios are vast complexes of buildings sprawling over an area as large as some towns. Universal City in California is the biggest. Covering 0.6 sq miles (1.7 sq km), the studio has its own fire department, zoo, stables, and police department.

Stunted development
Stuntmen replace regular actors for dangerous sequences. Stunts are filmed from a distance and later edited carefully. Stuntmen often stood in for female stars in the past, but laws have made this illegal in the United States. A studio can employ a stuntman in drag only when all available stuntwomen have refused the part.

Wet weekend
To simulate rivers, lakes, and oceans, set builders create vast water-filled tanks. The largest tank ever built filled an unused German airship hangar. Constructed in 1929, the 2,000-ft (600-m) long tank represented Russia's Volga River.

That sinking feeling
Like the small squid in the model shop, this giant is a latex model. It's a kind of huge puppet, realistic only on the side that faces the cameras. This doesn't mean it's harmless: the rubber shark in *Jaws* (US 1975, prod. William S. Gilmore, Jr.) accidentally sank the ship carrying the cast and crew, sending the camera to the seabed.

Stirring up a storm
To simulate a hurricane, movie-makers fix an aircraft propeller to a powerful electric motor. Besides a convincing storm, the vast fan creates a lot of noise, and a realistic soundtrack must be added later.

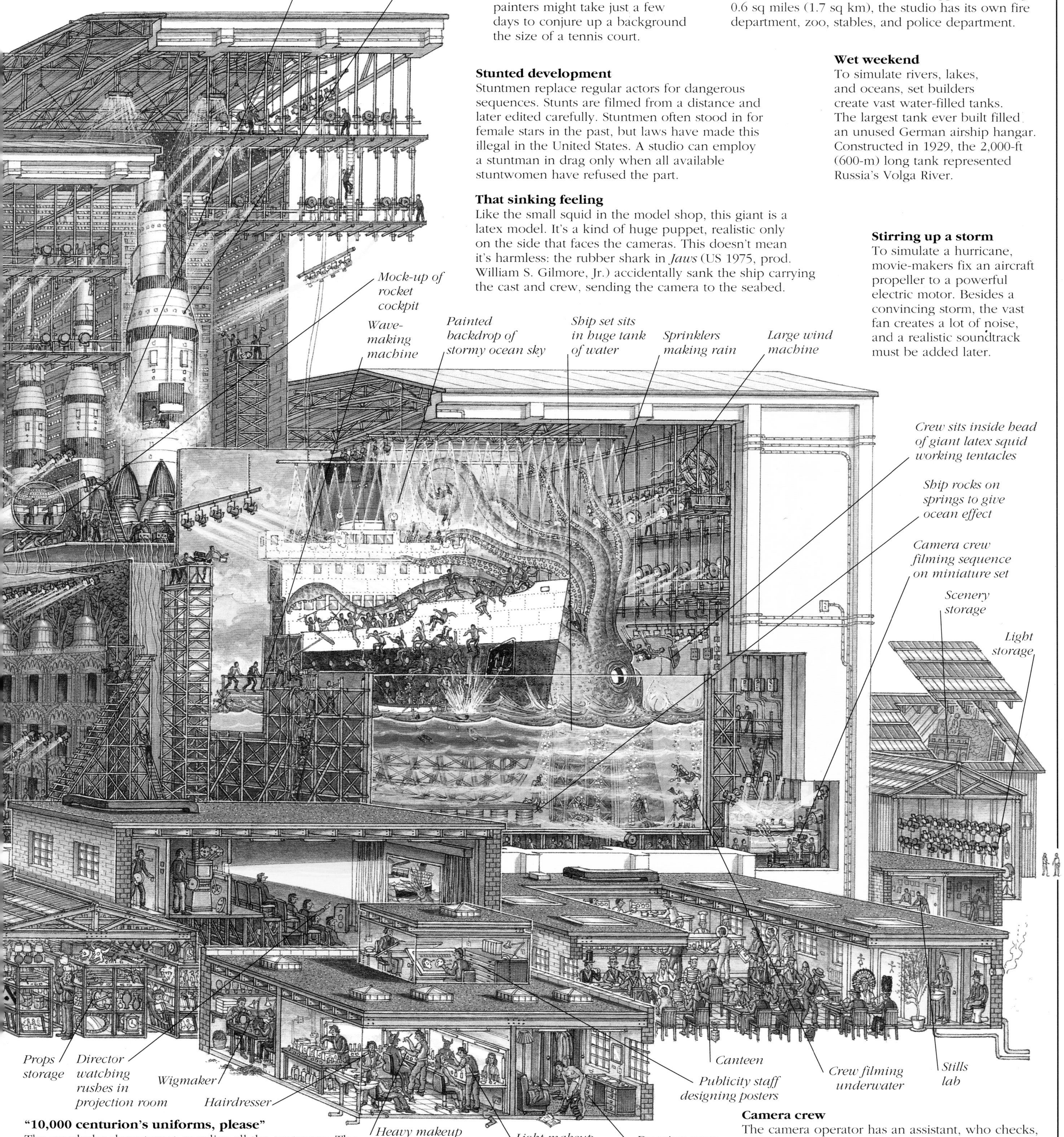

Mock-up of rocket cockpit

Wave-making machine

Painted backdrop of stormy ocean sky

Ship set sits in huge tank of water

Sprinklers making rain

Large wind machine

Crew sits inside head of giant latex squid working tentacles

Ship rocks on springs to give ocean effect

Camera crew filming sequence on miniature set

Scenery storage

Light storage

Props storage

Director watching rushes in projection room

Wigmaker

Hairdresser

Canteen

Publicity staff designing posters

Crew filming underwater

Stills lab

Heavy makeup (the devil)

Light makeup (stunning beauty)

Dressing room

"10,000 centurion's uniforms, please"
The wardrobe department supplies all the costumes. The audience sees major stars close-up, so the costumes they wear must be authentic in every detail. Crowd scenes require vast numbers of costumes: for the 1951 movie *Quo vadis*, the wardrobe department kept track of 29,000 outfits.

First look
Each studio has a tiny projection room of its own. There the director views the "rushes" – quick prints of all the scenes shot the previous day.

Camera crew
The camera operator has an assistant, who checks, cleans, and maintains the camera and focuses the lens. The clapper-loader marks the start and end of each scene by clapping the clapperboard and is responsible for loading the film. The key grip on a movie is responsible for moving the camera.

VENICE

IMAGINE A CITY IN THE SEA. HALF SINKING, HALF FLOATING, ITS ROOFS AND GLITTERING towers rise mysteriously from swirling mist. This magic place was Venice in the 17th century. Already 11 centuries old, the city had begun as a mud-bank refuge from war on Italy's mainland. Protected from attack by its shallow lagoon, Venice grew into a great world power. An exclusive club of noblemen led by the Doge (duke) ruled Venice ruthlessly. They became wealthy by controlling overland trade between Asia and Europe. The discovery of a sea route to Asia ended the monopoly in 1488; but despite the centuries of decline that followed, the city's fabulous beauty and legendary art treasures survived almost unchanged. The picture below shows the central section of the city.

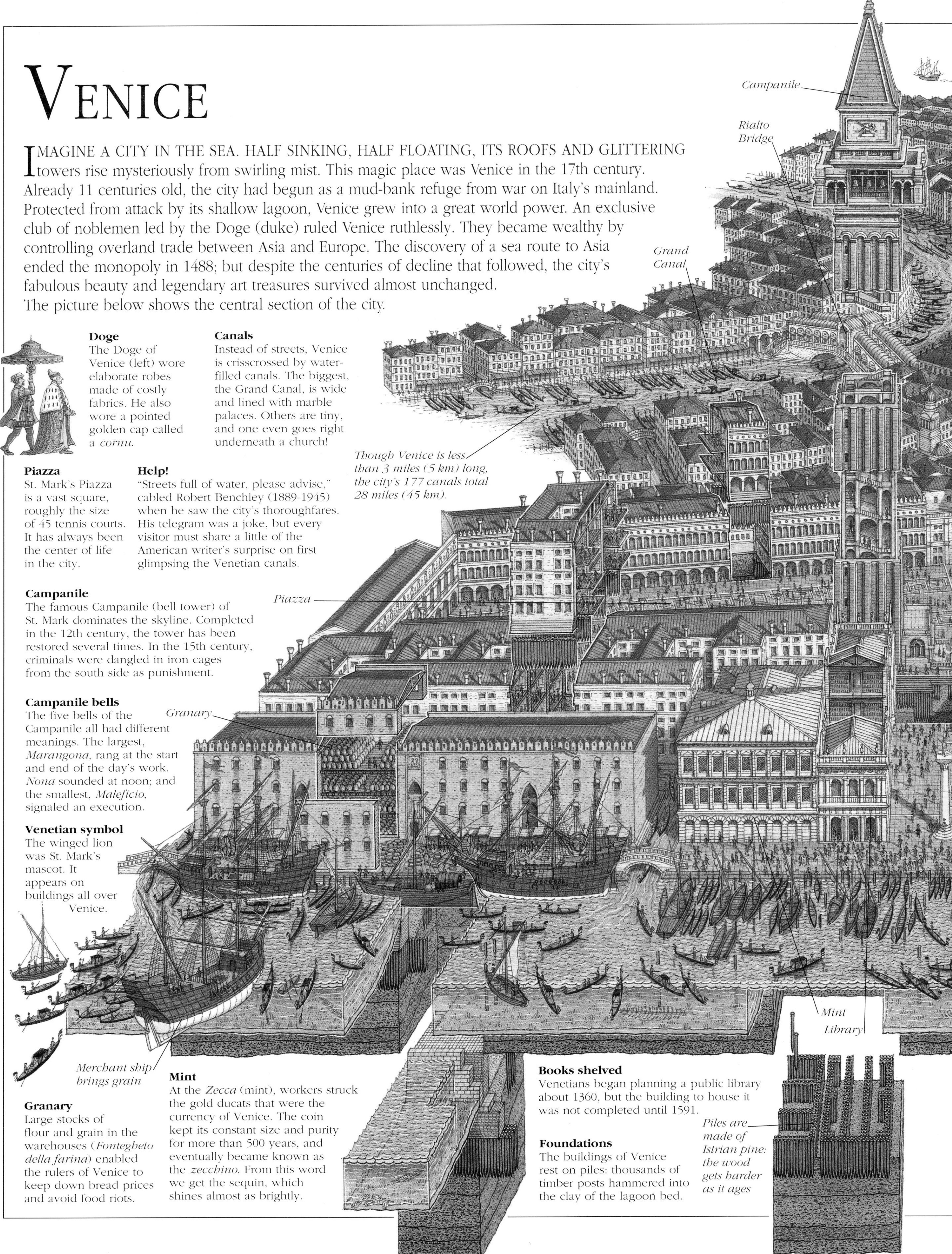

Doge
The Doge of Venice (left) wore elaborate robes made of costly fabrics. He also wore a pointed golden cap called a *cornu*.

Canals
Instead of streets, Venice is crisscrossed by water-filled canals. The biggest, the Grand Canal, is wide and lined with marble palaces. Others are tiny, and one even goes right underneath a church!

Piazza
St. Mark's Piazza is a vast square, roughly the size of 45 tennis courts. It has always been the center of life in the city.

Help!
"Streets full of water, please advise," cabled Robert Benchley (1889-1945) when he saw the city's thoroughfares. His telegram was a joke, but every visitor must share a little of the American writer's surprise on first glimpsing the Venetian canals.

Campanile
The famous Campanile (bell tower) of St. Mark dominates the skyline. Completed in the 12th century, the tower has been restored several times. In the 15th century, criminals were dangled in iron cages from the south side as punishment.

Campanile bells
The five bells of the Campanile all had different meanings. The largest, *Marangona*, rang at the start and end of the day's work. *Nona* sounded at noon; and the smallest, *Maleficio*, signaled an execution.

Venetian symbol
The winged lion was St. Mark's mascot. It appears on buildings all over Venice.

Granary
Large stocks of flour and grain in the warehouses (*Fontegheto della farina*) enabled the rulers of Venice to keep down bread prices and avoid food riots.

Mint
At the *Zecca* (mint), workers struck the gold ducats that were the currency of Venice. The coin kept its constant size and purity for more than 500 years, and eventually became known as the *zecchino*. From this word we get the sequin, which shines almost as brightly.

Books shelved
Venetians began planning a public library about 1360, but the building to house it was not completed until 1591.

Foundations
The buildings of Venice rest on piles: thousands of timber posts hammered into the clay of the lagoon bed.

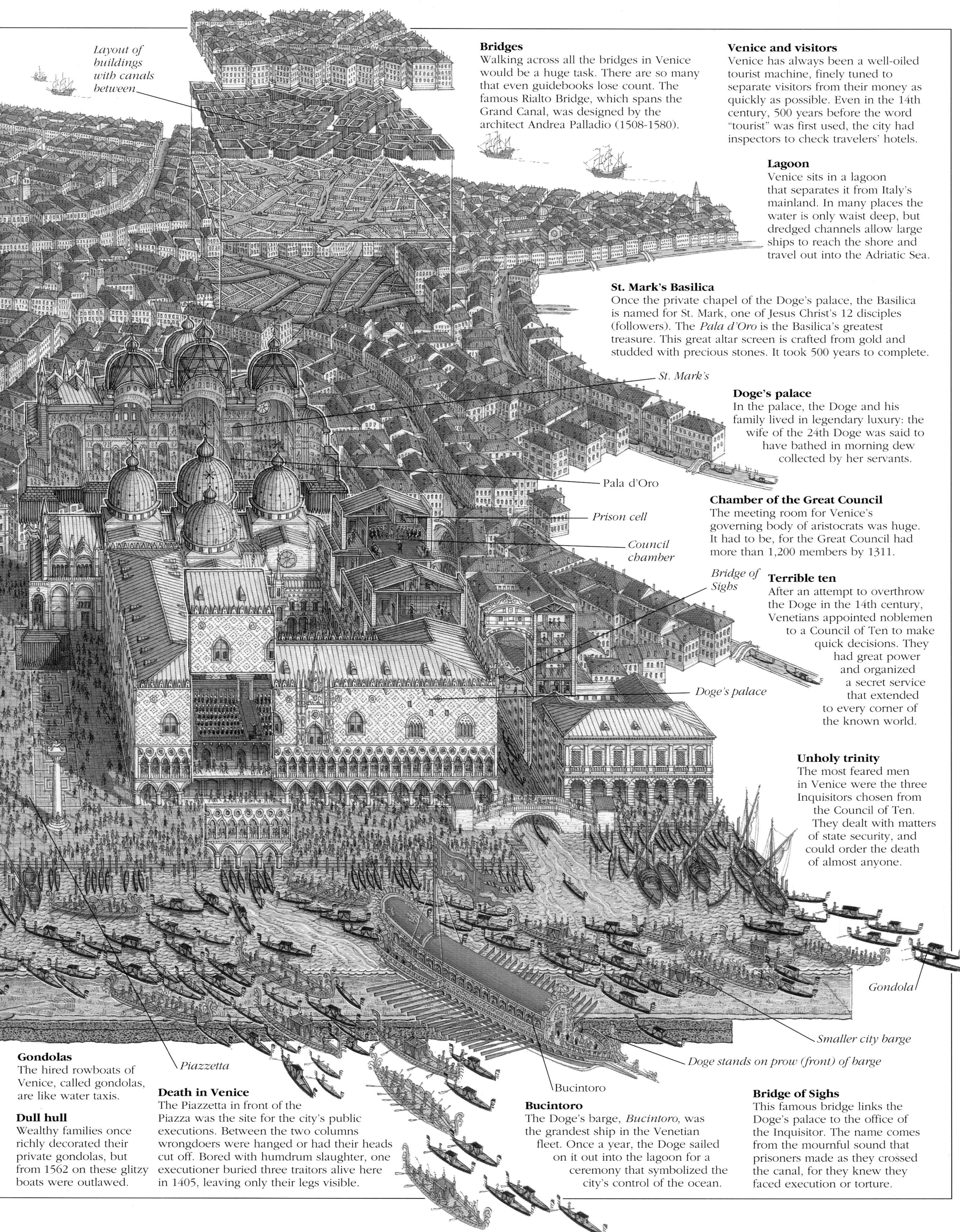

Bridges
Walking across all the bridges in Venice would be a huge task. There are so many that even guidebooks lose count. The famous Rialto Bridge, which spans the Grand Canal, was designed by the architect Andrea Palladio (1508-1580).

Venice and visitors
Venice has always been a well-oiled tourist machine, finely tuned to separate visitors from their money as quickly as possible. Even in the 14th century, 500 years before the word "tourist" was first used, the city had inspectors to check travelers' hotels.

Lagoon
Venice sits in a lagoon that separates it from Italy's mainland. In many places the water is only waist deep, but dredged channels allow large ships to reach the shore and travel out into the Adriatic Sea.

St. Mark's Basilica
Once the private chapel of the Doge's palace, the Basilica is named for St. Mark, one of Jesus Christ's 12 disciples (followers). The *Pala d'Oro* is the Basilica's greatest treasure. This great altar screen is crafted from gold and studded with precious stones. It took 500 years to complete.

Doge's palace
In the palace, the Doge and his family lived in legendary luxury: the wife of the 24th Doge was said to have bathed in morning dew collected by her servants.

Chamber of the Great Council
The meeting room for Venice's governing body of aristocrats was huge. It had to be, for the Great Council had more than 1,200 members by 1311.

Terrible ten
After an attempt to overthrow the Doge in the 14th century, Venetians appointed noblemen to a Council of Ten to make quick decisions. They had great power and organized a secret service that extended to every corner of the known world.

Unholy trinity
The most feared men in Venice were the three Inquisitors chosen from the Council of Ten. They dealt with matters of state security, and could order the death of almost anyone.

Gondolas
The hired rowboats of Venice, called gondolas, are like water taxis.

Dull hull
Wealthy families once richly decorated their private gondolas, but from 1562 on these glitzy boats were outlawed.

Death in Venice
The Piazzetta in front of the Piazza was the site for the city's public executions. Between the two columns wrongdoers were hanged or had their heads cut off. Bored with humdrum slaughter, one executioner buried three traitors alive here in 1405, leaving only their legs visible.

Bucintoro
The Doge's barge, *Bucintoro*, was the grandest ship in the Venetian fleet. Once a year, the Doge sailed on it out into the lagoon for a ceremony that symbolized the city's control of the ocean.

Bridge of Sighs
This famous bridge links the Doge's palace to the office of the Inquisitor. The name comes from the mournful sound that prisoners made as they crossed the canal, for they knew they faced execution or torture.

Tower Bridge

LONDON'S LANDMARK BRIDGE LOOKS LIKE A TALLER TWIN OF THE city's famous Tower nearby. Yet for all the pinnacles and parapets, the castlelike masonry is only skin deep. Under its stone cladding the bridge has a skeleton of steel. Built at the end of the 19th century to relieve traffic congestion, Tower Bridge is a bascule bridge. When a tall ship sails up London's Thames River, the two bascules, or leaves that carry the roadway, lift to allow the ship to pass underneath.

Heavy see-saw!
The word *bascule* means "see-saw" in French. On Tower Bridge, lead and steel weights counterbalance the bascules. The weights each weigh 320 tons (325 tonnes) – as much as 40 African elephants.

Weighty wind
Each bascule, or leaf, weighs 1,345 tons (1,220 tonnes) and is 100 ft (30 m) long. On a still day, little force is needed to lift the bridge. Yet the bridge can still open against the strongest gale-force gusts – equivalent to lifting each leaf with 150 cars parked on it.

Tower of strength
The original specification for the bridge required that it should be capable of being armed with guns.

Original paint work
Painters covered the metalwork of the bridge with three coats of "bright chocolate" paint before opening day.

Royal opening
The bridge had a royal opening in 1894. Excited crowds gazed as the bridge opened. But not everything went according to plan. Amplified sound was still 20 years in the future, so the royal opening speeches were completely inaudible more than a few feet from the platform.

Chain links
The bridges that link the piers to the banks on either side are suspension bridges; so the supports for the roadway are called "chains," despite the fact that they are really girders.

Spiral staircase access to roof space

Elevator winding gear

Heraldic achievement (crest)

Bascule

High level walkway

Bascule locking bolt

Gothic revival Scottish baronial architecture

Suspension chains

Interior steel framework

Anchor tie to north approach road

North approach road

Gas lamp

Elevator carries 25 passengers to high level walkway

Stairs to first floor

Control cabin

Nighttime signal light

Levers to operate bridge

Anchor tie inside concrete block

Entrance to the Tower of London

Archway leading to "Dead Man's Hole"

Pier clad in Cornish granite

Gault brickwork in cement

Machinery chamber contains two lifting engines

Passageway to accumulator chamber

Plunger

Lead counterweight

Pig iron weights

Dead Man's Hole
An archway under the north approach road gave access to "Dead Man's Hole" – a temporary morgue for bodies fished out of the river close to the bridge.

Built on mud
The bridge foundations stand on clay, not on bedrock. To prevent the bridge from sinking, the foundations are huge. The engineers kept the pressure down to less than 4 tons per sq ft (4.3 kg per sq cm) – less than the pressure of someone standing on your toe.

Bridgemaster's perks
The bridge employed 80 people, 14 on watch at any one time, including a Superintendent Engineer and a Bridgemaster, who had an official residence nearby.

Open wide
The bridge had to provide a very wide opening so that vast square-rigged sailing ships could pass through easily under wind power.

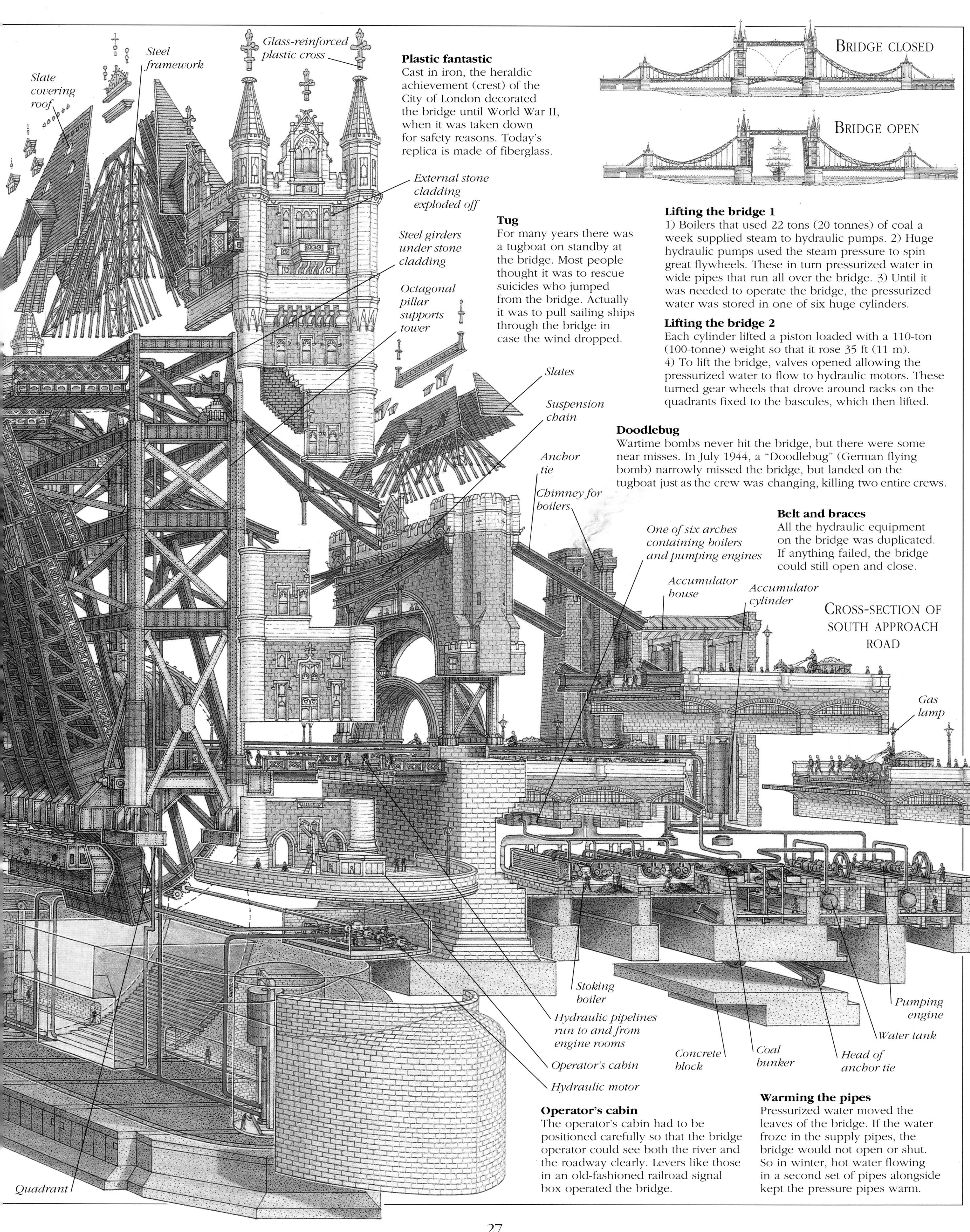

Plastic fantastic
Cast in iron, the heraldic achievement (crest) of the City of London decorated the bridge until World War II, when it was taken down for safety reasons. Today's replica is made of fiberglass.

Tug
For many years there was a tugboat on standby at the bridge. Most people thought it was to rescue suicides who jumped from the bridge. Actually it was to pull sailing ships through the bridge in case the wind dropped.

Lifting the bridge 1
1) Boilers that used 22 tons (20 tonnes) of coal a week supplied steam to hydraulic pumps. 2) Huge hydraulic pumps used the steam pressure to spin great flywheels. These in turn pressurized water in wide pipes that run all over the bridge. 3) Until it was needed to operate the bridge, the pressurized water was stored in one of six huge cylinders.

Lifting the bridge 2
Each cylinder lifted a piston loaded with a 110-ton (100-tonne) weight so that it rose 35 ft (11 m). 4) To lift the bridge, valves opened allowing the pressurized water to flow to hydraulic motors. These turned gear wheels that drove around racks on the quadrants fixed to the bascules, which then lifted.

Doodlebug
Wartime bombs never hit the bridge, but there were some near misses. In July 1944, a "Doodlebug" (German flying bomb) narrowly missed the bridge, but landed on the tugboat just as the crew was changing, killing two entire crews.

Belt and braces
All the hydraulic equipment on the bridge was duplicated. If anything failed, the bridge could still open and close.

Operator's cabin
The operator's cabin had to be positioned carefully so that the bridge operator could see both the river and the roadway clearly. Levers like those in an old-fashioned railroad signal box operated the bridge.

Warming the pipes
Pressurized water moved the leaves of the bridge. If the water froze in the supply pipes, the bridge would not open or shut. So in winter, hot water flowing in a second set of pipes alongside kept the pressure pipes warm.

HUMAN BODY

A TRIP THROUGH THE HUMAN BODY WOULD BE LIKE SURFING THROUGH THE VAST network of subways and sewers beneath a city. From the pounding heart, you'd slow gradually as the wide tunnels grew narrower and lower, before cruising back to repeat the ride just a minute or so later. A trip around the lungs would be far quicker – sucked down the ribbed trachea, you'd float along increasingly narrow tubes. Just a second later, hurricane-force winds would blow you out again. The brain controls the whole network, sending vital messages sparking through a web of tiny nerves.

Masticating muscle
The masseter muscle enables us to bite and is the body's strongest muscle. Normal bite pressure is about 165-220 lb (75-100 kg), but some people have a bite strength of nearly half a ton.

Reliable ticker
The human heart is an efficient and hard-working blood pump. It's only about the size of a fist, yet it lasts more than 70 years, pumping 30 times its own weight of blood a minute.

Not really bored
The yawn reflex draws air into the lungs when the brain needs more oxygen. This often happens after a long period of inactivity.

Blood
Our bodies contain about 1.3 gallons (5 liters) of blood – roughly half a bucketful. Three-fifths of it is coursing through our veins, while at any one moment our lungs contain about two and a half cups of blood. The rest is in the heart, arteries, and capillaries.

Lymphatic system
Another set of vessels in our body carries the lymphatic system, which circulates a thin, milky liquid called lymph. A line of defense and a source of nutrients, lymph contains disease-fighting white blood cells that combat infection. Filters in lymph vessels are known as glands, or nodes. When you are ill, germ-killing cells collect in the nodes to help you fight back.

Gray lump
Despite the awesome power of the brain, it's not much to look at. Some compare the brain's function with a computer's, but the human brain is many millions of times more powerful than even the biggest computers.

Hearing bones
Tiny bones in the ear amplify the vibrations of the eardrum to enable us to hear quiet sounds. The smallest bone, the stapes or stirrup bone, is the smallest in the human body. It weighs 0.07-0.15 oz (2-4.3 g), about as much as a sugar cube.

No bones about it
The "funny bone" is actually the ulnar nerve, positioned next to the humerus.

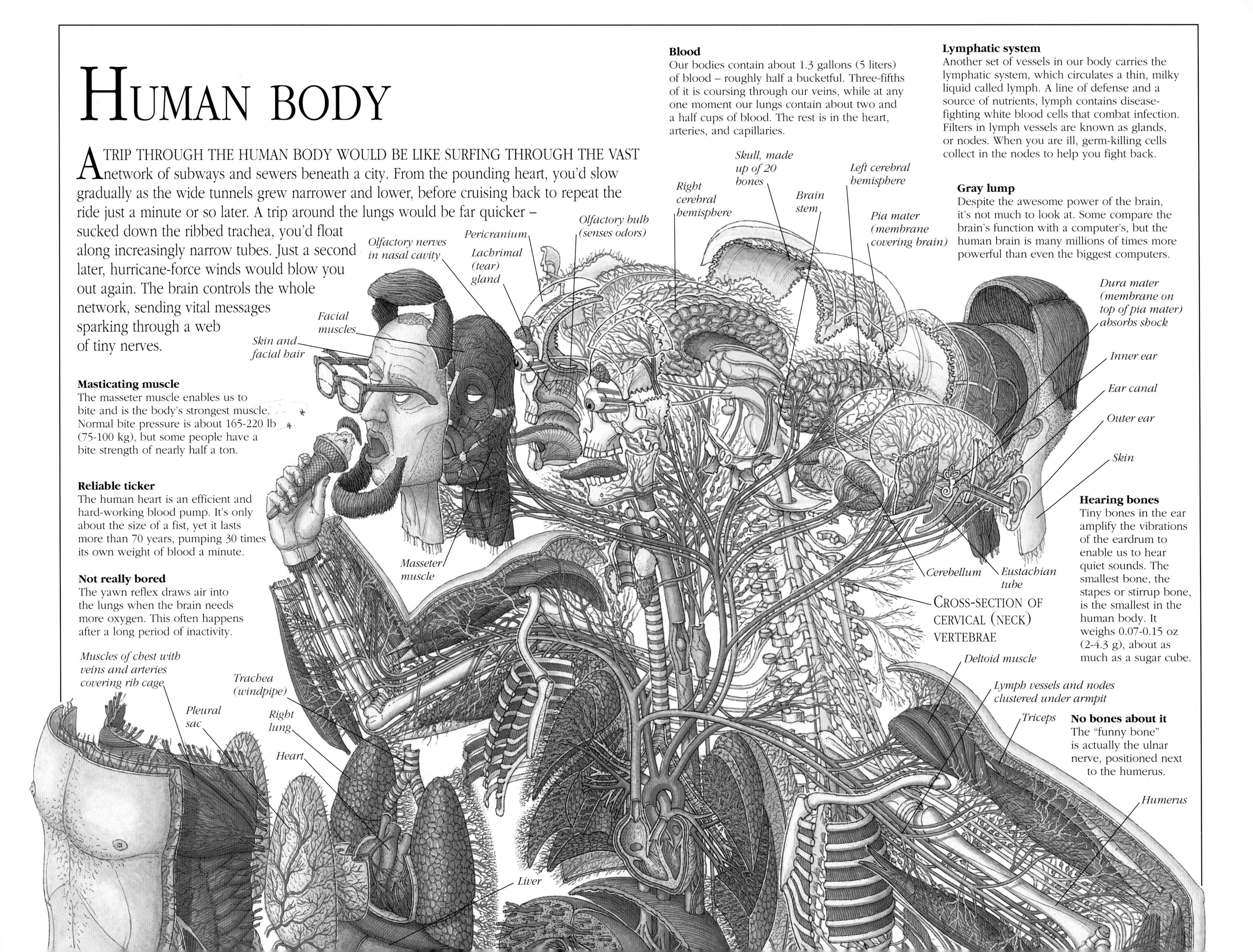

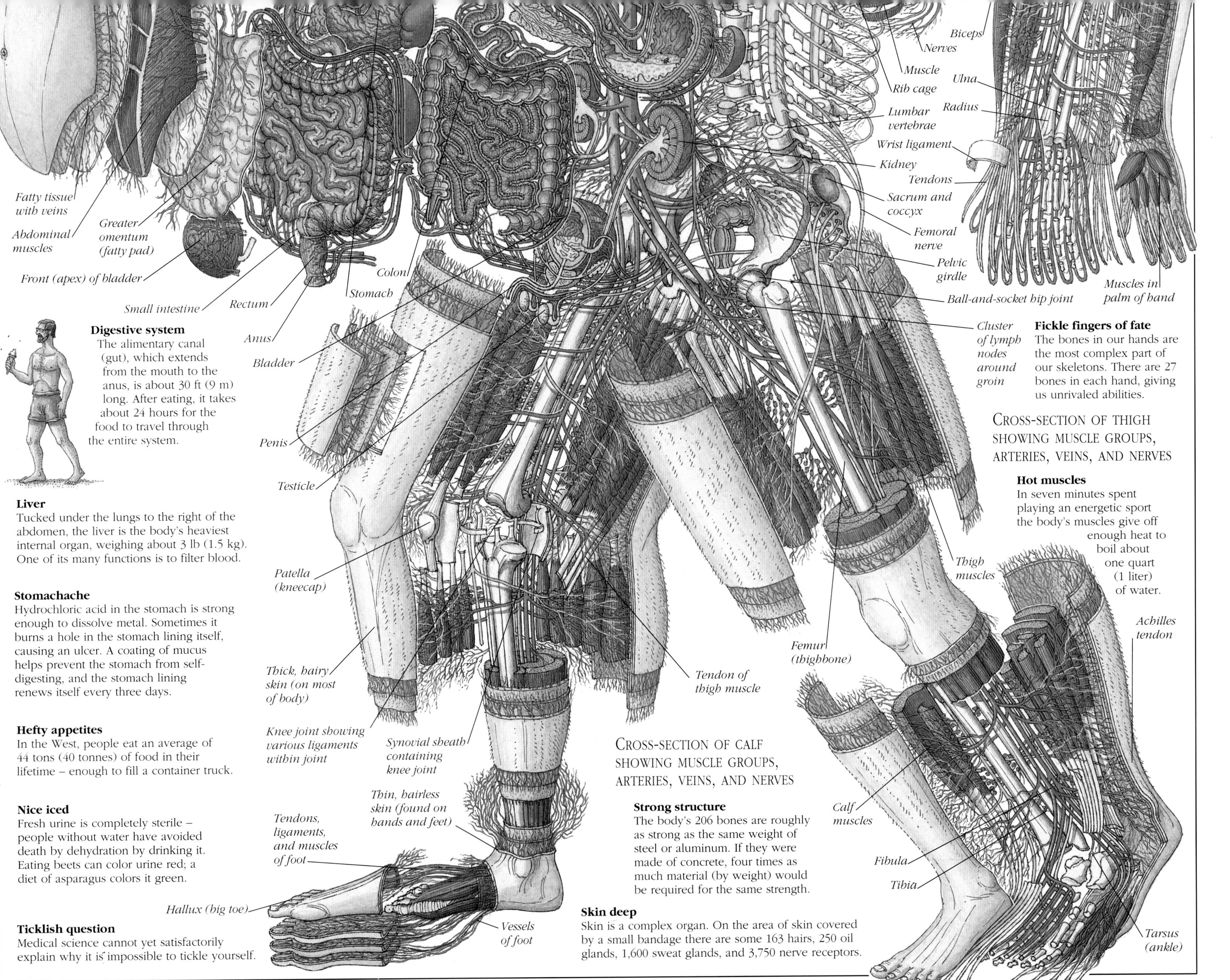

CROSS-SECTION OF THIGH SHOWING MUSCLE GROUPS, ARTERIES, VEINS, AND NERVES

CROSS-SECTION OF CALF SHOWING MUSCLE GROUPS, ARTERIES, VEINS, AND NERVES

Digestive system
The alimentary canal (gut), which extends from the mouth to the anus, is about 30 ft (9 m) long. After eating, it takes about 24 hours for the food to travel through the entire system.

Liver
Tucked under the lungs to the right of the abdomen, the liver is the body's heaviest internal organ, weighing about 3 lb (1.5 kg). One of its many functions is to filter blood.

Stomachache
Hydrochloric acid in the stomach is strong enough to dissolve metal. Sometimes it burns a hole in the stomach lining itself, causing an ulcer. A coating of mucus helps prevent the stomach from self-digesting, and the stomach lining renews itself every three days.

Hefty appetites
In the West, people eat an average of 44 tons (40 tonnes) of food in their lifetime – enough to fill a container truck.

Nice iced
Fresh urine is completely sterile – people without water have avoided death by dehydration by drinking it. Eating beets can color urine red; a diet of asparagus colors it green.

Ticklish question
Medical science cannot yet satisfactorily explain why it is impossible to tickle yourself.

Fickle fingers of fate
The bones in our hands are the most complex part of our skeletons. There are 27 bones in each hand, giving us unrivaled abilities.

Hot muscles
In seven minutes spent playing an energetic sport the body's muscles give off enough heat to boil about one quart (1 liter) of water.

Strong structure
The body's 206 bones are roughly as strong as the same weight of steel or aluminum. If they were made of concrete, four times as much material (by weight) would be required for the same strength.

Skin deep
Skin is a complex organ. On the area of skin covered by a small bandage there are some 163 hairs, 250 oil glands, 1,600 sweat glands, and 3,750 nerve receptors.

Grand Canyon

Each year, nearly five million people flock to the southwestern United States to see ... nothing! They marvel at a vast chunk of emptiness 10 miles (15 km) wide and more than a mile (1,500 m) deep. Of course, tourists really come to see what the huge gap reveals: a candy-striped fantasy of layered rock. The raging Colorado River created this spectacular gash when it carved its way through the rocks of Arizona, taking ten million years to complete its work. Just as amazing as the layer cake geology is the climate and wildlife of the canyon. The base is a desert where rocks get hot enough to fry eggs in the summer sun. Yet high above on the rim there are pine forests where mountain lions hunt deep in freezing winter snows.

Count 'em
One of the first Europeans to see the canyon wrote: "Ours has been the first, and will doubtless be the last, party of whites to visit this profitless locality." Now nearly five million tourists visit each year!

Cliffs and slopes
Hard rocks form vertical cliffs; softer rocks weather into steep slopes.

Splat!
Anybody falling from the rim of the canyon would soon strike the sloping sides, but if they had a clear drop to the bottom, they'd have about 35 seconds to admire the view on the way down.

No snow
Snow often dusts the canyon's high North Rim, but it almost never falls to the foot of the canyon.

Crossing the canyon
Backpackers crossing the canyon usually start on the higher and more remote North Rim. The rim-to-rim record is less than four hours, but most take 2-4 days.

Exhausted
Some hikers can't complete the climb; the park rangers carry them to the top on mules. These "drag-outs" must pay their fare before getting into the saddle.

That's deep
The Grand Canyon is deep enough in places to stack four Empire State Buildings one on top of the other. Its maximum width at the base is equivalent to five ocean liners lined up bow-to-stern.

Climatic zones

Boreal (forest) life zone
The highest point in the canyon is the North Rim, parts of which are more than 8,140 ft (2,480 m) above sea level. The aspen, fir, and spruce trees are adapted to shed the heavy snows that fall each year.

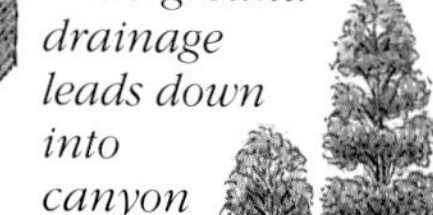

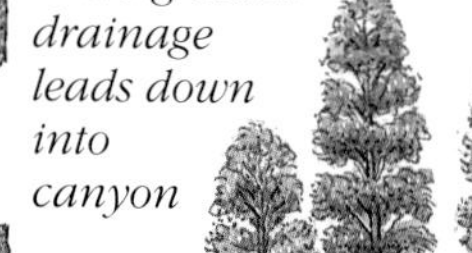

Transition zone
Along the rim in parts of the canyon is a transition zone. Its climate and wildlife resemble those of both the Boreal zone above and the Upper Sonoran zone below.

Upper Sonoran zone
The life zone of the upper side of the canyon is called the Upper Sonoran zone. At the top, juniper and piñon pine trees dominate.

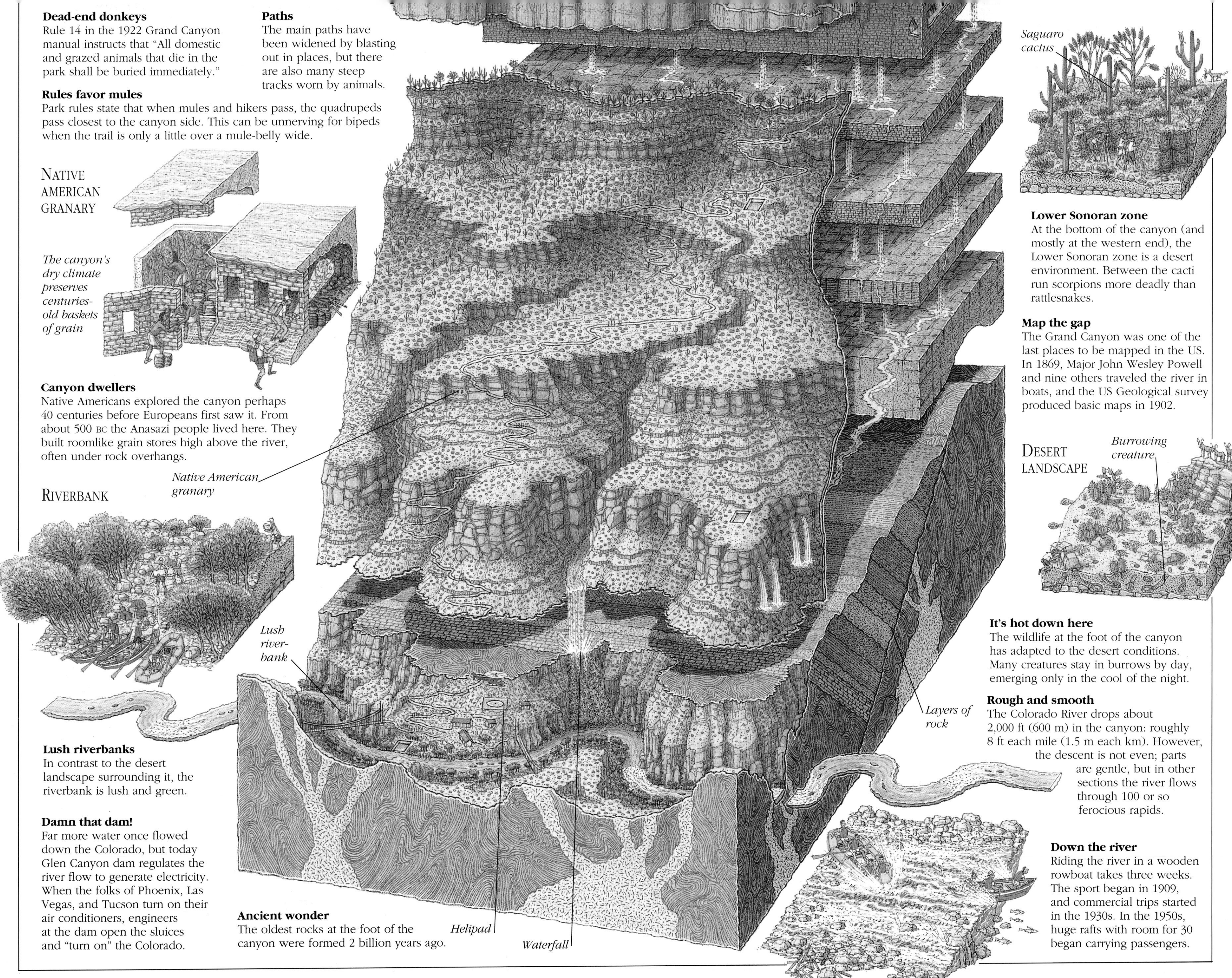

Dead-end donkeys
Rule 14 in the 1922 Grand Canyon manual instructs that "All domestic and grazed animals that die in the park shall be buried immediately."

Paths
The main paths have been widened by blasting out in places, but there are also many steep tracks worn by animals.

Rules favor mules
Park rules state that when mules and hikers pass, the quadrupeds pass closest to the canyon side. This can be unnerving for bipeds when the trail is only a little over a mule-belly wide.

Canyon dwellers
Native Americans explored the canyon perhaps 40 centuries before Europeans first saw it. From about 500 BC the Anasazi people lived here. They built roomlike grain stores high above the river, often under rock overhangs.

Lush riverbanks
In contrast to the desert landscape surrounding it, the riverbank is lush and green.

Damn that dam!
Far more water once flowed down the Colorado, but today Glen Canyon dam regulates the river flow to generate electricity. When the folks of Phoenix, Las Vegas, and Tucson turn on their air conditioners, engineers at the dam open the sluices and "turn on" the Colorado.

Ancient wonder
The oldest rocks at the foot of the canyon were formed 2 billion years ago.

Lower Sonoran zone
At the bottom of the canyon (and mostly at the western end), the Lower Sonoran zone is a desert environment. Between the cacti run scorpions more deadly than rattlesnakes.

Map the gap
The Grand Canyon was one of the last places to be mapped in the US. In 1869, Major John Wesley Powell and nine others traveled the river in boats, and the US Geological survey produced basic maps in 1902.

It's hot down here
The wildlife at the foot of the canyon has adapted to the desert conditions. Many creatures stay in burrows by day, emerging only in the cool of the night.

Rough and smooth
The Colorado River drops about 2,000 ft (600 m) in the canyon: roughly 8 ft each mile (1.5 m each km). However, the descent is not even; parts are gentle, but in other sections the river flows through 100 or so ferocious rapids.

Down the river
Riding the river in a wooden rowboat takes three weeks. The sport began in 1909, and commercial trips started in the 1930s. In the 1950s, huge rafts with room for 30 began carrying passengers.

INDEX

ACKNOWLEDGMENTS

DK would like to thank the following people who helped in the preparation of this book:

Lynn Bresler for the index
Constance Novis for editorial support
Bohdan Paraschak for research

B.A.A. plc, Heathrow
Jack Fryer and the Cranbrook Windmill Association
Kent Fire Safety Division
Shelter, The National Campaign for Homeless People
Shepperton Studios for access to soundstages and workshops
Special thanks to:
Lt. Katherine A. McNitt, Station Chief, National Oceanographic and Atmospheric Administration,
Amundsen-Scott South Pole Station